T0295312

Phycoremediation Processes in Industrial Wastewater Treatment

Increasing population and industrialization are the key pollutant contributors in water bodies. The wastes generated by industries are highly hazardous for humans and the ecosystem and require a comprehensive and effective treatment before being discharged into water bodies. Over the years, many up gradations have been introduced in traditional water treatment methods which were expensive and ineffective especially for removal of toxic pollutants. Phycoremediation has been gaining attention due to its mutual benefit in wastewater treatment and for valuable algae biomass production. Wastewater, especially sewage and industrial effluents, is rich in pathogenic organisms, organic and inorganic compounds and heavy metals that adversely affect human and aquatic life. Microalgae use these inorganic compounds and heavy metals for their growth. In addition, they also reduce pathogenic organisms and release oxygen to be used by bacteria for decomposition of organic compounds in a secondary treatment. In this book, the potential of microalgae in wastewater treatment, their benefits, strategies, and challenges are discussed. The increasing need of finding innovative, low-cost, low-energy, sustainable and eco-friendly solutions for wastewater treatment makes the publication of a book on phycoremediation timely and appropriate.

Features:

1. Deals with the most emerging aspects of algal research with special reference to phycoremediation.
2. Studies in depth diversity, mutations, genomics and metagenomics study
3. An eco-physiology, culturing, microalgae for food and feed, biofuel production, harvesting of microalgae, separation and purification of biochemicals.

Wastewater Treatment and Research
Series Editor: Maulin P. Shah

Wastewater Treatment: Molecular Tools, Techniques, and Applications
Maulin P. Shah

Advanced Oxidation Processes for Wastewater Treatment: An Innovative Approach
Maulin P. Shah, Sweta Parimita Bera, and Gunay Yildiz Tore

Emerging Technologies in Wastewater Treatment
Maulin P. Shah

Bio-Nano Filtration in Industrial Effluent Treatment: Advanced and Innovative Approaches
Maulin P. Shah

Membrane and Membrane-Based Processes for Wastewater Treatment
Maulin P. Shah

Phycoremediation Processes in Industrial Wastewater Treatment
Maulin P. Shah

For more information, please visit: https://www.routledge.com/Wastewater-Treatment-and-Research/book-series/WASTEWATER

Phycoremediation Processes in Industrial Wastewater Treatment

Edited by
Maulin P. Shah

CRC Press is an imprint of the
Taylor & Francis Group, an **informa** business

First edition published 2023
by CRC Press
6000 Broken Sound Parkway NW, Suite 300, Boca Raton, FL 33487-2742

and by CRC Press
4 Park Square, Milton Park, Abingdon, Oxon, OX14 4RN

CRC Press is an imprint of Taylor & Francis Group, LLC

Library of Congress Cataloguing-in-Publication Data
Names: Shah, Maulin P., editor.
Title: Phycoremediation processes in industrial wastewater treatment / edited by Maulin P. Shah.
Description: First edition. | Boca Raton : CRC Press, [2023] |
Series: Wastewater treatment and research | Includes bibliographical references.
Identifiers: LCCN 2022055717 (print) | LCCN 2022055718 (ebook) |
Subjects: LCSH: Phycoremediation. | Sewage--Purification.
Classification: LCC TD756.8 .P46 2023 (print) | LCC TD756.8 (ebook) |
DDC 628.3--dc23/eng/20230130
LC record available at https://lccn.loc.gov/2022055717
LC ebook record available at https://lccn.loc.gov/2022055718

ISBN: 978-0-367-76007-6 (hbk)
ISBN: 978-0-367-76010-6 (pbk)
ISBN: 978-1-003-16510-1 (ebk)

DOI: 10.1201/9781003165101

Typeset in Times
by MPS Limited, Dehradun

Contents

Editor Bio

Maulin P. Shah is very interested in genetic adaptation processes in bacteria, the mechanisms by which they deal with toxic substances, how they react to pollution in general and how we can apply microbial processes in a useful way (like bacterial bioreporters). One of his major interests is to study how bacteria evolve and adapt to use organic pollutants as novel growth substrates. Bacteria with new degradation capabilities are often selected in polluted environments and have accumulated small (mutations) and large genetic changes (transpositions, recombination, and horizontally transferred elements). His work has been focused to assess the impact of industrial pollution on microbial diversity of wastewater following cultivation dependant and cultivation independent analysis. He has had more than 280 research publications, and is an Editorial Board member for many highly-reputed national and international journals; he has edited 20 special issues on Industrial Wastewater Treatment and Research themes, in high-impact journals with publishers Elsevier, Springer, Wiley, and Taylor & Francis; as well as editing 75 books in wastewater microbiology and industrial wastewater treatment.

Contributors

Abudukeremu Kadier
Laboratory of Environmental Science and Technology
The Xinjiang Technical Institute of Physics and Chemistry
Key Laboratory of Functional Materials and Devices for Special Environments
Chinese Academy of Sciences
Urumqi, China

Arvind Singh
Department of Chemical Engineering & Biochemical Engineering
Rajiv Gandhi Institute of Petroleum Technology Jais
Amethi, Uttar Pradesh, India

Bekti Marlena
Center of Industrial Pollution Prevention Technology
The Ministry of Industry Republic of Indonesia
Semarang, Central Java, Indonesia

Chiranjib Bhattacharjee
Chemical Engineering Department
Jadavpur University
Kolkata, West Bengal, India

Divya
Department of Biotechnology
Sir J. C. Bose Technical Campus
Kumaun University Nainital
Bhimtal, Uttarakhand, India

Hiren K. Patel
School of Science
P. P. Savani University
Surat, Gujarat, India

Jaya Lakkakula
Amity Institute of Biotechnology
Amity University Maharashtra
Mumbai, Maharashtra, India

Jaydeep Dobariya
School of Science
P. P. Savani University
Surat, Gujarat, India

Khalida Bloch
Department of Microbiology
School of Science
RK University
Rajkot, Gujarat, India

Khushboo Dasauni
Department of Biotechnology
Sir J. C. Bose Technical Campus
Kumaun University Nainital
Bhimtal, Uttarakhand, India

Mohamad Rusdi Hidayat
Institute for Industrial Research and Standarization of Pontianak
The Ministry of Industry Republic of Indonesia
Pontianak, West Kalimantan, Indonesia

Muhilan Mahendhiran
Unidad Multidisciplinaria de Docencia e Investigación
Facultad de Ciencias
Universidad Nacional Autónoma de México (UNAM)
Puerto de abrigo s/n, Sisal
Yucatán, México

and

Kalasalingam School of Agriculture and Horticulture (KSAH)
Kalasalingam Academy of Research and Education (Deemed to be University)
Krishnankovil, Tamil Nadu, India

Nandini Girish
Department of Bioanalytical Sciences
Ramnarain Ruia Autonomous College
Mumbai, Maharashtra, India

Nilesh S. Wagh
Amity Institute of Biotechnology
Amity University Maharashtra
Mumbai, Maharashtra, India

Peng Cheng Ma
Laboratory of Environmental Science and Technology
The Xinjiang Technical Institute of Physics and Chemistry
Key Laboratory of Functional Materials and Devices for Special Environments
Chinese Academy of Sciences
Urumqi, China

Perumalla Srikanth
Kalasalingam School of Agriculture and Horticulture
Kalasalingam Academy of Research and Education
Krishankoil, Tamil Nadu, India

M.R. Rajan
Department of Zoology
Micheal job College of Arts and Science for Women
Coimbatore Dist, Tamil Nadu, India

Ranjana Das
Chemical Engineering Department
Jadavpur University
Kolkata, West Bengal, India

Rishee K. Kalaria
Aspee Shakilam Biotechnology Institute
Navsari Agricultural University
Surat, Gujarat, India

Rupak Kishor
Department of Chemical Engineering
Maulana Azad National Institute of Technology
Bhopal, Madhya Pradesh, India

Rustiana Yuliasni
Center of Industrial Pollution Prevention Technology
The Ministry of Industry Republic of Indonesia
Semarang, Central Java, Indonesia

Sachin Palekar
Department of Bioanalytical Sciences
Ramnarain Ruia Autonomous College
Mumbai, Maharashtra, India

Sandhya Menon
Department of Bioanalytical Sciences
Ramnarain Ruia Autonomous College
Mumbai, Maharashtra, India

Sanjay Kumar
Department of Chemical Engineering
Faculty of Technology
Marwadi University
Rajkot, Gujarat, India

Setyo Budi Kurniawan
Department of Chemical and Process Engineering
Faculty of Engineering and Built Environment
National University of Malaysia (UKM)
Bangi, Selangor, Malaysia

Shankar Durairaj
K.M. College of Pharmacy
Uthangudi, Madurai, Tamil Nadu, India

Sivakumar Durairaj
Kalasalingam School of Agriculture and Horticulture
Kalasalingam Academy of Research and Education
Krishankoil, Tamil Nadu, India

Sougata Ghosh
Department of Microbiology
School of Science
RK University
Rajkot, Gujarat, India

D. Suganya
Department of Biology
Gandhigram Rural Institute - Deemed University
Dindigul Dist, Tamil Nadu, India

Tapan K. Nailwal
Department of Biotechnology
Sir J. C. Bose Technical Campus
Kumaun University Nainital
Bhimtal, Uttarakhand, India

Vijay Laxmi Yadav
Indian Institute of Technology
Banaras Hindu University (BHU)
Varanasi, Uttar Pradesh, India

1 Addressing the Strategies of Algal Biomass Production with Wastewater Treatment

Divya, Khushboo Dasauni, and Tapan K. Nailwal
Department of Biotechnology, Sir J. C. Bose Technical Campus, Kumaun University Nainital, Bhimtal, Uttarakhand, India

CONTENTS

1.1 INTRODUCTION

Since the accessibility of freshwater assets is declining due to anthropogenic activities, reuse of water is widely recommended for effective wastewater

DOI: 10.1201/9781003165101-1

management. The sustainable management of water is needed today to create a healthier and self-sufficient environment. To do so, three key driving forces, economic efficiency, innovation, and novelty, must be implemented. Discharge of wastewater without proper treatment leads to the depletion of the aquatic ecosystem due to eutrophication. To treat wastewater, both developed and developing nations are implementing strategies based on the biological mode (Han et al., 2012). Microalgae are ubiquitous and versatile photoautotrophic microorganisms. They own the ability to adapt readily under diverse conditions (Diniz et al., 2017). Phycoremediation 'the microalgal technology' was initiated by Oswald in the 1950s. Commercial cultivation of algae on wastewater also offers the opportunity for production of value-added products, for example, biofuels (Han et al., 2015). The widely used technologies for algal cultivation with wastewater systems that are also available on the commercial scale, include open pond cultivation systems:

- raceway pond,
- high rate algal ponds,
- Bioreactors with light energy, often called photobioreactors (PBRs).

Because the cost involved in the production of media to culture microalgae is very high, a wastewater resource has motivated microalgal farming. This approach could simultaneously afford both bioremediation and biomass production. Algae can be cultivated in nutrient and organic-rich wastewater from agriculture, aquaculture, and municipal wastewater to remediate excess nutrients from effluents. On the other hand, considering that heavy metals pose a great danger to the ecosystem and human health due to their everlasting toxicity, their high concentration in wastewater is a concern. Microalgae's special affinity for metals enables the removal of hazardous metals effectively. Due to the bioaccumulation effect of microalgae cultivated in metal-polluted wastewater, they uptake a certain concentration of heavy metals. Wastewater treatment via microalgae cultivation is considered to be a promising pathway to produce biofuel because microalgae have the capability of recovering nitrogen, phosphorus, organic compounds, and even heavy metals in wastewater (Cai et al., 2013; Chan et al., 2014; Renuka et al., 2015).

1.2 BIOLOGY OF MICROALGAE

The microscopic organisms, microalgae, are made up of eukaryotic cells driven by the same photosynthetic process as higher plants. Microalgal cells encompass the cell wall, plasma membrane, cytoplasm, nucleus, and organelles. Microalgae have chlorophyll-containing plastids, responsible for carrying out photosynthesis. Unlike higher plants, every cell in microalgae is photoautotrophic with the ability to directly absorb nutrients. In chloroplastm the cells of microalgae photons assimilate as energy and CO_2 extracted from exhaust gases generated by combustion process or bacterial respiration; nutrients from wastewater are used to synthesize their biomass concurrently producing oxygen. The transformation of CO_2 and water into organic compounds demands no extra energy addition, which helps prevent secondary pollution. The liberated oxygen from microalgae is sufficient to achieve the desired aerobic requirement of bacteria for promoting the metabolic breakdown of the residual organic substances present in treated wastewater. Moreover, microalgae also need a light and dark regime for photosynthesis, where the former is employed for a photochemical phase that leads to the generation of adenosine triphosphate (ATP) and nicotinamide adenine dinucleotide phosphate-oxidase (NADPH), while the latter is for the biochemical phase that leads to production of essential biomolecules for growth.

1.3 BIOREMEDIATION OF WASTEWATER AND CULTIVATION OF MICROALGAE

The biomass generated by microalgae in wastewater treatment can be extracted and converted into a variety of products such as value-added chemicals via biorefinery and fuel bioproducts (Chew et al., 2017) and bio-oil production via thermochemical conversion, biochar via pyrolysis, biomethane through anaerobic digestion, and biodiesel through trans-esterification (Klinthong et al., 2015; Yu et al., 2018). Biomass is also considered imperative for the removal of heavy metals from wastewater as it performs passive biosorption process as an uptake mechanism. The uptake mechanism involves a biosynthesis process to form biomass by photosynthetic microalgae together with uptake of ammonium, phosphate, nitrate found in wastewater.

$$CO_2 + H_2O + NH_4 + NO_3 + PO_4 + \text{Photons (light)} + \text{Algae}$$
$$= C_{106}H_{263}O_{110}N_{16}P + C_6H_{12}O_6 + O_2$$

On other hand, heterotrophic microalgae are well known to consume organic carbon, soluble carbonates dissociated from carbon acids either by direct intake or by converting CO_2 to carbonates through enzymatic activity of carbonic anhydrase. Therefore, heterotrophic microalgae using the provided organic carbon substrate $[CH_2O]_n$ as the sole carbon source can omit light energy and replace the synthesis process to form organic carbon by absorbing CO_2 during photosynthesis. In addition to providing nutrients, the production of algal biomass also requires light energy. Stoichiometric formula for the most common components in microalgae cells is $C_{106}H_{181}O_{45}N_{16}P$. These elements should be present in the culture medium to achieve the optimal growth of microalgae and successfully eradicate inorganic compounds from wastewater. In large-scale experiments, many methods have been reported for culturing microalgae with wastewater to remove nutrients and increase biomass productivity, especially methods using pond systems and photo-bioreactor (PBR), which have proven effective for the growth of microalgae (Oswald et al., 1988). However, different strains of microalgae have different efficiencies in removing nutrients from these wastewaters and producing biomass. Obviously, compared with photobioreactors, productivity of waterway pond is much higher, although the latter requires more doubling time. For example, a 90% dilution of organic-rich swine wastewater can support microalgae growth and lipid productivity (6.3 mg l^{-1} d^{-1}) (Wang et al., 2012).

Similarly, diluting or pre-treating high-strength wastewater can improve growth and production of algae (Woertz et al., 2009; Kong et al., 2010). Although the lead content in wastewater containing sodium glutamate is high and acidic, diluting it to 25% can promote the growth of algae (Miao et al., 2016). Similarly, highly diluted metal-rich wastewater produced 0.37 g l^{-1} of biomass to support the growth of microalgae (Ji et al., 2016). The presence of nitrogen and phosphorus in wastewater plays a crucial role in the metabolism of algae. When nitrogen or phosphate is restricted and deprived, growth is inhibited. Therefore, limiting substrates must be determined to enhance growth of algae and thus, contribute to WWT. For example, the significant difference in N:P ratio in wastewater caused a decrease in biomass productivity in urban wastewater (Posadas et al., 2015). In some cases, the use of high-strength wastewater to cultivate microalgae can inhibit growth due to the presence of inorganic chemicals and high chemical oxygen demand (COD) (Cuellar-Bermudez et al., 2017; Min et al., 2011). The influence of light on the cultivation of microalgae is also essential to achieve the dual benefits, namely, the realization of the biomass production rate and wastewater remediation. Obviously, the light intensity of PBR in the pond system is very high (Gupta et al., 2015). However, the design principle of PBR with high volume ratio and intense mixing can ensure the uniformity of illumination of all batteries, and it is essential to maximize the utilization of PAR. As the depth of pond increases, paddle-wheel mixing and turbulent eddies passing through the water flow will help the cells to be evenly exposed to light (Park et al., 2011). Removal of nutrients from wastewater is low due to limited light as a consequence of its impact on respiration (Aslan et al., 2006). In general, nutrient removal kinetics is given by the formula:

$$P_i = P_0 \times e^{-kt}$$

where, P_i is the nutrient concentration at time t_i and P_0 is the initial concentration of the nutrient, and k is the first order reaction constant. Thus, the biomass yield (Y) linked with the consumption of nutrients can be calculated using the formula:

$$Y = X - X_0/P_0 - P$$

where, P is the concentration of nutrients when there is no significant decrease of nutrients, X is the concentration of Chl a with respect to the nutrient P, and X_0 is the initial concentration of Chl a. The removal of pollutants was efficient in the case of diurnal cycles rather than under continuous illumination (Lee and Lee, 2001). On the pilot scale, the use of high light intensity may lead to higher biomass productivity and nutrient removal, but it often leads to photo inhibition and destruction of photoreceptors (Park et al., 2011). Restricted conditions can increase biomass productivity. In addition to preventing ammonia through volatilization and phosphorus through precipitation, the supply of CO_2 also helps to adjust the optimal pH value so that ammonia can be absorbed (Park et al., 2011). However, the transfer of CO_2 is an important factor to be considered during cultivation because CO_2 easily escapes into the air. In addition, adding CO_2 from flue gas instead of pure CO_2 can increase the nutrient removal rate.

1.4 APPROACHES FOR ENHANCING THE MICROALGAL CULTIVATION

1.4.1 Microalgal-Bacterial Energy Nexus

The microalgae-bacterial energy connection is one of the most important strategies. It can remove pollutants from wastewater through a mutually beneficial method, that is, using CO_2 generated during bacterial respiration as a carbon source for microalgae and using microalgae photosynthesis. Oxygen produced in the process is utilized for energy yield by bacteria. However, bacteria release extracellular substances and enzymes that are toxic to microalgae (Abinandan et al., 2015; Torres et al., 2017). Similarly, assimilation of complex organic carbon sources present in wastewater by microalgae is very stringent, and sometimes, the chances of growth inhibition are very high (Ramakrishnan et al., 2010). The interaction of algae and bacteria also plays an important role in flocculation process. This may be due to factors such as cell-surface charge and secretion of extracellular polymers by bacteria (Gutzeit et al., 2005). Nevertheless, addition of bacteria in the nexus helps to transform complex constituents into simple organic substances that can easily be assimilated by microalgae (Lee and Lee, 2001). For example, the growth of *Chlorella pyrenoidosa* in soybean wastewater with bacteria will cause the bacteria to quickly decompose complex organic matter, thereby producing readily available organic components for microalgae growth (Zhang et al., 2012). Under light conditions, microalgae can produce oxygen and other metabolites. The concoction of diverse bacterial species act synergistically for the removal of pollutants. Similarly, through the co-growth of *Chlorella* and *Bacillus licheniformis*, 78% of NH_4 and 92% of total phosphate (TP) originally contained in wastewater were removed (Liang et al., 2013).

In addition, the biomass produced can be used in energy-conversion systems, such as anaerobic digestion or microbial fuel cells (MFC). Anaerobic environment in fuel cells can help produce hydrogen, and the oxygen produced can be used by bacteria without affecting the growth of algae (White et al., 2009). Gajda et al. (2015) demonstrated the simultaneous reduction of waste and energy production through bacteria-algae system. A brief review by Luo et al. (2017) focused on algae and microbial communities to integrate wastewater treatment and energy potential through fuel cells.

1.4.2 Microalgal Cultivation Strategies

To obtain higher biomass productivity to produce value-added products, a proper cultivation strategy is essential. To achieve it, microalga are cultivated photoautotrophically, photoheterotrophically, or mixotrophically (Brennan and Owende, 2010). Although there have been some recent innovations to increase the biomass productivity of microalgae during WWT, the success of this technology for biomass production depends on the absorption capacity of microalgae nutrients in wastewater (Abinandan et al., 2015). This necessitates the use of new methods in microalgae culture to obtain dual benefits, namely remediation and biomass production (Osundeko and Pittman, 2014). In the recent past, different strategies have been adopted for

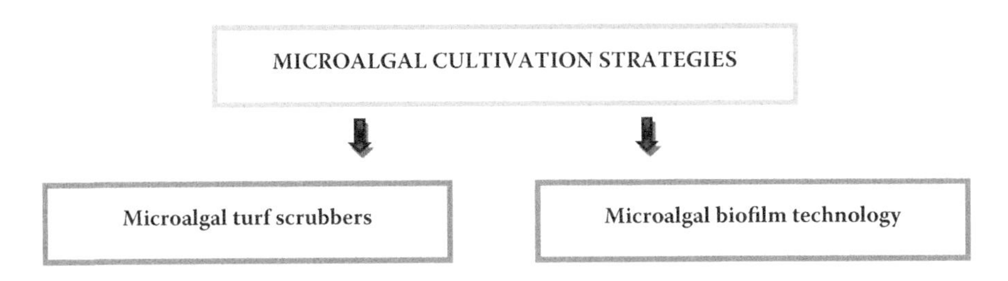

FIGURE 1.1 Strategies of microalgal cultivation.

wastewater treatment; these include biomass productivity and lipid biosynthesis, for example, beer wastewater (BWW) treatment by *Chlorella* spp. It involves a two-stage photoautotrophic-photoheterotrophic system in which the first mode (photoautotroph) uses digested BWW for algae cultivation for 7 days, and then the second mode (photoheterotroph) starts using the original BWW, which acts as a carbon source (Farooq et al., 2013). This strategy doubled biomass and increased yield (dry weight) of lipids by 30%. Lee et al. (2016) observed that algae bacterial culture effectively removed nutrients when cultivated in a two-stage photoperiod cycle (i.e., 12:60 h light:dark 6 d, then 12:12 h light:dark 8 d) in municipal wastewater. When substrate-limiting factors or stresses are present, the derived by-products can be used as an alternative organic carbon source for the growth of microalgae. For example, during WWT, the use of crude glycerol derived from biodiesel production in microalgae culture can enhance biomass production and bioremediation (Ma et al., 2016) (Figure 1.1).

1.4.2.1 Microalgal Biofilm Technology

Recently, application of algae biofilm technology has attracted people's attention as an alternative to suspension technology for the treatment of wastewater and the resulting biomass production. This is a natural phenomenon in which extracellular polysaccharides, proteins, and other metabolites are produced to help form algae biofilms (Berner et al., 2015). Production of algal biofilms requires a solid lower surface, which is usually a matrix layer composed of polymers (Schnurr and Allen, 2015). Singh et al. (2006) described the mechanism of biofilm formation with respect to time and other parameters. Extracellular polysaccharides and biofilm formation in microalgae play an important role (Delattre et al., 2016; Xiao and Zheng, 2016). Compared with suspended matrix technology, biofilms can be easily separated from wastewater. However, this requires development of specific design systems to interact with the cell surface and standard flow from the gradient of biomass. This is essential for production of biofilms (Qureshi et al., 2005).

For example, biofilms can fit well in materials with many textures to create additional areas that are generally suitable for WWT systems (Fitch et al., 1998). Since efficacy of the system depends on amount of light exposure to biofilm, some systems have been designed to allow maximum light transmittance. Biomass falls into the suspension, making the culture easy to harvest, thereby reducing harvesting costs (Schnurr et al., 2013). On different substrates, biofilms prepared from different strains of microalgae have different biomass production rates and pollutant removal rates from wastewater. While testing the efficiency of microalgal biofilms in the outdoors under various light conditions for phosphate removal with an operating time of 2 years, Sukačová et al. (2015) observed the rapid removal of phosphate under continuous illumination when compared to diurnal cycles and solar radiation. Boelee et al. (2014) pointed out that due to the limited light distribution on biofilms grown in low-concentration wastewater, addition of CO_2 has no effect on nutrient removal and biomass production.

Use of the rotating algae biofilm reactor (RABR) system was evaluated to reduce nutrients and harvest cells. Christenson and Sims (2012) used suspension technology to evaluate the efficacy of RABR in wastewater treatment and observed that the nutrients in RABR reduce biomass production due to the formation of biofilms and presence of residual biomass, even after cell harvest. Moreover, compared with traditional watercourse ponds, the RABR using watercourses has a 40% higher biomass production within 20 d of operation time. In addition, total suspended solids (TSS) reached 50% after 2 days and 90% within 4 days. Light loss was observed during batch cultivation, which affected the productivity of biomass.

1.4.2.2 Microalgal Turf Scrubbers

Algae turf scrubber (ATS) is an engineered algae system for WWT (Adey and Loveland, 2011). The wastewater is sent to a tank where algae cells can grow on the sieve placed in it (Mulbry et al., 2008). ATS method is associated with biofilm formation, which requires attachment materials for the growth of algae. Although ATS is a promising technology of WWT and requires minimal energy and operating investment, it still requires a large effort to improve efficiency and has different operating parameters (Pizarro et al., 2006). In a preliminary study, ATS reduced the nutrient content in Florida water, thereby removing pollutants such as PO_4, NO_3, NO_2, and NH_4, respectively, by 16%, 49%, 19%, and 41% (D'Aiuto et al., 2015).

When treating secondary wastewater with different hydraulic loading rates, Craggs et al. (1996) reported a drop in hydraulic power, which increased removal of pollutants. Similarly, using filamentous green algae to treat anaerobic-digested manure can increase biomass productivity, but it cannot remove nutrients and increase the sample loading rate. Mulbry et al. (2008) reported a significant improvement in nutrient removal from diluted dairy wastewater at higher loading rates. Obviously, ATS provides a positive approach in WWT, where energy consumption obviously depends on process design and performance.

1.4.2.3 Microalgal Cultivation Systems: Merits and Demerits

Obviously, extensive and in-depth research has been conducted to produce biofuels from algae, which makes algae biotechnology more promising and feasible (Brennan and Owende, 2010). However, these studies are mostly limited to laboratory or pilot scale, which is a limiting factor for implementation of algae technology on a global scale. Use of waterway ponds to improve biomass productivity, together with wastewater remediation, seems very promising (Mehrabadi et al., 2015). Other variables (such as light, pH, temperature, and dissolved oxygen) may affect productivity, but these factors can be overcome by increasing the pool depth, determining the exact retention time, and using PBR. However, due to well-organized net-energy consumption and maintenance work, it is not economically feasible to implement WWT on a large scale in a large PBR. Biofilm cultivation system, however, surpasses these disadvantages of this technology with reduced light restriction, gaseous mass transfer, and better water-use efficiency (Christenson and Sims, 2011). In addition, the efficiency of biofuel obtained from biomass in biofilm technology is limited. Use of ATS technology for WWT is reliable, but it still needs to improve its application in large-scale high-strength wastewater mass production. Latest trends in microalgae cultivation strategies seem to be promising for effective WWT.

Therefore, the algae-bacteria system for WWT is the best method for commercialization because it combines energy efficiency and effective remedial measures. Harvesting of microalgae cells/biomass is a challenging stage of phycoremediation technology, and new methods need to be explored to significantly improve the harvesting process. Current trends indicate that flocculation and sedimentation minimize the energy consumption of microalgae harvesting (Milledge and Heaven, 2013). Compared with the cultivation system used to harvest biomass, the biofilm technology shows a lower energy value. Therefore, no important reports on the application of biofilms in WWT systems for commercial-scale biomass harvesting are available. Use of flocculating microalgae to flocculate non-flocculating counterparts seems promising because it can increase the productivity of biomass, leading to higher biofuel production.

The mechanism behind heavy-metal biosorption has been systematically summarized previously, including metal absorption and release pathways through metabolism and non-metabolism-dependent processes during photosynthesis. In the process of controlling the cultivation and value-added process of wastewater fed with microalgae, the way of absorption and release of heavy metals is very important to control the morphology of the process and the final treatment technology. The mechanism of wastewater biosorption and bioremediation of microalgae has been extensively studied (Salama et al., 2017).

1.5 MULTIFACETED ROLE: MICROALGAL-BASED WASTEWATER TREATMENT

In many developing countries, the shortage of clean water supply may be due to failure to properly treat wastewater or discharge wastewater below environmental safety levels into nearby water bodies. In particular, water quality in overpopulated countries such as India, Kenya, Ethiopia, and Nigeria has reached a state of concern (Onuoha, 2012). Due to industrial, agricultural, and household activities, organic and inorganic impurities and various pollutants (from trace pollutants to heavy metals and excessive nutrients) are discharged into nearby water bodies (Rathod, 2014). Wastewater contains a large amount of organic and inorganic nutrients. Due to its high biological and chemical oxygen demand (BOD and COD), the ecosystem is out of balance. The presence of excessive nutrients such as nitrogen (N) and phosphorus (P) can lead to eutrophication of the water body, thereby destroying the health of the water system. Algae-based wastewater-treatment technologies provide compelling solutions due to their ability to effectively fix inorganic compounds including CO_2 and heavy metals (Koppel et al., 2018, 2019). Microalgae show strong absorption capacity of inorganic nutrients because they require nitrogen and phosphorus for protein synthesis and heavy metals used as micronutrients for growth (Chen et al., 2018; Liu et al., 2020). Thus, using algae as a wastewater bioremediation agent can effectively absorb nitrogen and phosphorus from wastewater, maintain dissolved oxygen content, and help reduce pathogens and fecal bacteria present in wastewater (Das et al., 2019). Observations obtained from the studies indicate that wastewater in contact with microalgae leads to a significant reduction in the content of heavy metals, nitrates, and phosphates (Rathod, 2014). Microalgae treatment is also a more efficient method of wastewater treatment because compared with conventional wastewater treatment, microalgae treatment can be treated in one step, while conventional wastewater treatment requires multiple processes to fix the ratio of carbon, nitrogen, and phosphorus (C: N: P). From an environmental point of view, it is also a sustainable choice because it has the ability to convert CO_2 into chemical substances and fuel products without causing pollution, which helps reduce greenhouse gas emissions.

1.5.1 Mechanism of Uptake

Using a microalgae-based system to treat wastewater can effectively remove inorganic compounds such as nitrates, phosphates, heavy metals, inorganic carbon, toxic substances (organic and inorganic), BOD, COD, and other impurities dissolved in wastewater through its absorption mechanism. Uptake and consumption of nitrate and phosphate by microalgae cells to promote growth can significantly reduce nitrogen and phosphorus content in wastewater and improve the quality of wastewater discharge (Emparan et al., 2019). Four species of microalgae were evaluated, namely S. *dimorphus*, *S. quadricauda*, *C. sorokiniana*, and *C. vulgaris* ESP-6 in membrane photobioreactor (MPBR) and ordinary photobioreactor (NPBR), and each type of microalgae was mixed with diluted wastewater after anaerobic digestion. Until the ninth day, when different microalgae strains were observed, the residual ammonia nitrogen and phosphate concentrations in MPBR and NPBR confirmed that the microalgae cultivated by MPBR can remove more nutrients than the NPBR microalgae. In another study, the ability of different microalgae strains (*Chlamydomonas* sp., *Chlorella* sp., and *Oocystis* sp.) on the removal of nitrogen and phosphate was studied by assessing NO_3-N and PO_4^{-3}-P loss (Rasoul-Amini et al., 2014). The removal rates of phosphate and nitrate were recorded every 4 days during the entire experiment. Results showed that all microalgae strains reached highest PO_4^{-3}-P removal efficiency on the last day of evaluation (day 14). This proves that a longer cultivation time will increase the amount of nutrients absorbed by microalgae. A series of experiments were conducted using *Chlorella vulgaris* to show the effect of temperature on the removal efficiency of nutrients, BOD, and COD of microalgae. The removal rates of total nitrogen and phosphorus by microalgae and growth rate of microalgae in 2 days were recorded. Results showed that after the temperature reached 30°C, the growth of algae and nutrient absorption began to decline rapidly. This shows that 30°C is the critical temperature in the experiment (Figure 1.2).

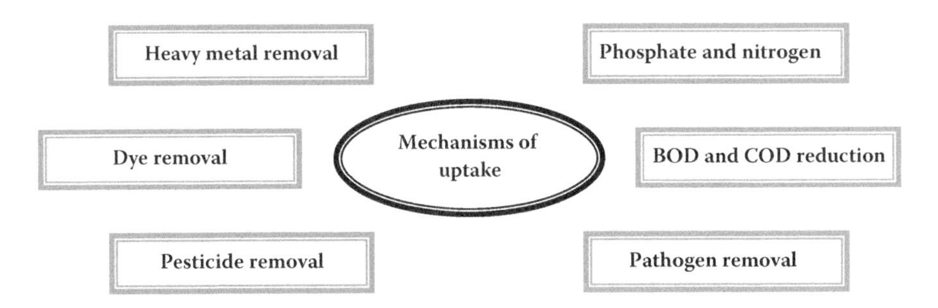

FIGURE 1.2 Microalgal uptake mechanism.

1.5.1.1 Phosphorus Uptake

Lipids, nucleic acids, and proteins naturally present in wastewater play an important role in energy metabolism and growth of microalgae. Transport of inorganic phosphate occurs on the plasma membrane of microalgae cells. During the metabolism of algae, inorganic phosphorus in the form of monophosphate and dihydrogen phosphate (HPO_4^{-2} and H_2PO_4) is incorporated into organic compounds through phosphorylation, in this case, adenosine diphosphate (ADP). The phosphorylation process requires energy to produce its final product, ATP. Energy can come from the oxidation of respiratory substrates, the mitochondrial electron transport system found in eukaryotic microalgae, and the light used in photosynthesis (Emparan et al., 2019). Cell walls and membranes of microalgae are usually composed of polysaccharides, proteins, and lipids. Membrane-transport proteins are essential to promote the interaction between microalgae and the concentrated heavy metal environment. The mechanism of heavy metal bioremediation includes non-metabolism-dependent uptake and metabolism-dependent uptake. Metabolism-independent processes mainly include cell-surface adsorption (ion exchange, complexation, physical adsorption, and precipitation) and extracellular adsorption (precipitation) (Salama et al., 2019). Precipitation process that depends on metabolism is usually affected by intracellular adsorption, and this adsorption is related to the transport of active metals across the membrane (He and Chen, 2014). Microalgae can easily use wastewater. During the metabolism of microalgae, ATP and glutamate are converted into ADP and PO_4^{-3} through phosphorylation. The removal rate of phosphorus in microalgae varies with the type of algae and wastewater (El-Sheekh et al., 2016). Microalgae may eliminate organic pollutants, such as phenolic compounds and surfactants, mainly through biosorption and enzymatic degradation (Xiong et al., 2016).

1.5.1.2 Nitrogen Uptake

Organic nitrogen can be discharged into wastewater through sewage from the land where animal manure is stored or applied. Organic nitrogen is a key element in biological substances such as enzymes, peptides, proteins, and chlorophyll and energy transfer molecules (such as ADP and ATP). Organic nitrogen comes from inorganic sources, including nitrite (NO_2), nitrate (NO_3), nitric acid (HNO_3), ammonia (NH_3), ammonium (NH_4^+), and nitrogen (N_2). Nitrogen in wastewater usually exists in the form of NH_4^+, NO_2, and NO_3. Eukaryotic microalgae can convert inorganic nitrogen into organic form through assimilation. In short, the transformation mechanism that occurs on the plasma membrane of microalgae is the reduction of nitrate (NO_3) to nitrite (NO_2) and ammonium (NH_4^+), which are then integrated into amino acids (the organic form of nitrogen). The first step of nitrate assimilation involves nitrate reductase (NR), which is the reduced form of NADH, alpha-diphosphopyridine nucleotide ($C_{21}H_{27}N_7O_{14}P_2$), which exists in microalgae and can transfer two electrons. Consequently, ferrodoxin (Fd) from microalgae along with nitrite reductase, which is NADPH, $C_{21}H_{29}N_7O_{17}P_3$ produced by photosynthesis involving ADP, phosphate, and NADP, are involved in the reaction of reducing NO_2 to NH_4. Six electrons are transferred in the process. Through this action, all inorganic nitrogen in the intracellular fluid of microalgae is reduced to NH_4, while phosphorylation releases ATP that binds ammonium to the amino acid (glutamine) in the intracellular fluid of microalgae (Emparan et al., 2019).

1.5.1.3 BOD and COD Reduction

Microorganisms in the wastewater consume dissolved oxygen released by algae, thereby degrading organic matter into CO_2 and water during the interaction between algae and bacteria. This process proves that microalgae can provide a large amount of molecular oxygen as an oxidant for bacterial oxidation, thereby reducing BOD and COD in wastewater. Results showed that, compared with the control treatment, textile wastewater microalgae inoculation significantly reduced the values of BOD and COD. Except for *Anabaena*, most microalgae have a relatively high removal efficiency of BOD and COD (>80%) (Lee et al., 2016).

The effect of temperature on the removal of nutrients, biochemical oxygen demand, and chemical oxygen demand of microalgae was also studied. The results showed that the removal percentage of COD and BOD continued to increase and reached a peak at 30°C. Therefore, the critical temperature is about 30°C, and any further increase in temperature will result in a decline in the growth rate of algae and a reduction in oxygen demand removal efficiencies. Therefore, less molecular oxygen will be released, resulting in an increase in BOD and COD value (Azeez, 2010).

1.5.1.4 Pathogen Removal

The pathogen-removal mechanism of microalgae in wastewater includes nutrient competition, increased pH and dissolved oxygen levels, pathogen adhesion and precipitation, and algal toxins (Dar et al., 2019). In microalgae culture, microalgae absorb nutrients and carbon sources, which are the main energy source for bacterial cells. Nutritional competition between microalgae and bacteria will result in bacterial cell starvation, and ultimately, lead to bacterial cell death. Due to CO_2 assimilation during photosynthesis, pH usually increases during the cultivation of microalgae. Absorption of nitrogen by microalgae also increases the pH of the medium because each nitrate ion is reduced to ammonia and produces an OH ion. This phenomenon leads to elimination of pathogens. Due to limited transfer of CO_2 from the atmosphere and oxygenation of microorganisms, microalgae will further increase the pH level, which may lead to the death of pathogens. Fluctuations in pH are also believed to have an adverse effect on the survival of *E. coli*, and, therefore, will significantly eliminate coliforms in feces, such as *E. coli, Enterococcus* and *Clostridium perfringens* (Ansa et al., 2011).

Adhesion of fecal bacteria to microalgae in wastewater is essential because when the pH rises and the dissolved oxygen produced by microalgae rises, it can ensure that bacterial cells are in close proximity. For adhesion to occur, the pathogen will first attach to the solid matter, which will sink as a precipitate and deposit on the surface of microalgae cells. Subsequently, the available polysaccharide expressed by a bacterial cell will form a positively charged amino group. Positively charged polymers will neutralize the surface of negatively charged microalgae, which will form a bridge between the particles, causing bacterial cells to adhere to microalgae (Dar et al., 2019). Microcystin LR toxin produced by algae strain *Synechocystis* sp and the long-chain fatty acid toxin produced by *Clostridium vulgaris* under high pH conditions have been found to be harmful to fecal bacteria and pathogens. *Chlorella* can remove coliforms in feces by secreting and increasing the level of chlorophyll-a (Ansa et al., 2012).

1.5.1.5 Pesticide Removal

Microalgae can absorb various organic pollutants, including pesticides, as energy sources for their growth in wastewater through biosorption and biodegradation. Biosorption includes the mechanisms of absorption, adsorption, surface complexation, ion exchange, and precipitation that occur in the cell walls of living and dead cells. Biodegradation occurs when enzymes produced by microalgae break down bonds in pesticide molecules. In a study conducted, *Chlorella vulgaris* was challenged with four common fungicides: propamocarb, mandipropamid, cyprodinil, and metalaxyl in two experiments: short-term include biosorption for 60 min and long-term involves biodegradation for 4 days. Cyprodinil pesticide was removed effectively, with lowest remaining pesticide for both short and long term, followed by mandipropamid. Another short-term and long-term experiment was conducted to analyze the percentage of pesticide removal by microalgae (Hussein et al., 2016). The microalgae used in the experiment was *Chlorella,* and the insecticides were oxalate, simazine, isoproturon, atrazine, propanol, carbofuran, dimethoate, pendimethalin, cresol, and humus. In a short-term experiment, a sample containing the corresponding pesticide and sterile Milli Q water (initial concentration 2 mg/l) and *Chlorella* (0.6 g dry weight/liter) were mixed at room temperature and stirred at 380 rpm for 1 h. In the long-term experiment, the pesticide samples were mixed with sterile BG11 (initial concentration of 2 mg/L) and *Chlorella* was grown on it (inoculation volume of 10% (v/v)) for 5 days. Results show that when compared with the short-term experiment, the microalgae pesticide removal effect in the long-term experiment was better.

1.5.1.6 Dye Removal

Microalgae are used to remove color and vinyl sulfone dyes from textile wastewater due to their high surface area and binding affinity (Andrade and Andrade, 2018; Chu and Phang, 2019). In the mechanism of dye removal, cell wall of microalgae involves biosorption, electrostatic attraction, complexation and biotransformation. Dye ions adhere and accumulate on the surface of algae biopolymer, and then diffuse to the solid phase of the biopolymer. Extracellular polymers containing functional groups will help the dye molecules bioadsorb to polymer surface (Kumar et al., 2014).

Spirulina is the biomass of microalgae species and has been proven to be an effective biosorbent for removal of reactive dyes. Biomass of haricot bean and scalp bean can remove basic dyes through biosorption. In addition, *Clostridium vulgaris* is also commonly used as a bioactive adsorbent to remove reactive dyes, such as Remazol Black B (Aksu and Tezer, 2005).

1.5.1.7 Heavy Metal Uptake

Due to the high risk of toxic heavy metals present in wastewater, a large amount of wastewater discharged from industry, agriculture, municipal, and animal factories poses a huge threat to the environment. If the concentration of heavy metals exceeds the allowable discharge limit for aquatic environment, the discharge of wastewater into the environment is also harmful to public health. Conventional heavy metal treatment technologies usually include ion exchange, lime precipitation, chemical precipitation, and adsorption. Actual use of these technologies is usually inefficient, imperfect, costly, and energy-efficient, especially for large amounts of wastewater with low concentrations of heavy metals (Jacinto et al., 2009). Compared with traditional techniques, growing microalgae to remove heavy metals has many unique advantages, including rapid metal absorption capacity of microalgae, energy saving potential, eco-friendly use, lower implementation costs, and the ability to absorb high and low concentrations of metals. Compared with traditional treatment technologies, the significant advantages of microalgae make it an effective candidate for bioremediation of polluted wastewater. Various microalgae (*Chlorella, Spirulina maxima, Spirulina, Coelastrum* sp, etc.) have the ability to remove heavy metals. According to reports, the total removal rates of heavy metals in the studied acid mine drainage wastewater were: Fe (94.89%), Cu (95.60%), Zn (94.19%), Mn (89.22%), As (87.50%), and Cd (95.00%), respectively (Choi, 2015). The results show that these functional groups play a vital role in metal biosorption. In addition, various plasma membrane metal transport proteins, such as ZIP, CTR, HMA1/HMA6, are responsible for the transfer of Zn^{+2}, Cu^{+2} and Cu^{+} into the cytoplasm (Salama et al., 2017). Arsenic is one of the most toxic metals in polluted wastewater. Microalgae (such as *Chlorella*) can efficiently absorb As (V) and As (III) through phosphate transporter, hydroglyceroglycoprotein, and hexose permease (Zhang et al., 2012) (Table 1.1).

1.6 COMMERCIAL APPLICABILITY OF THE MICROALGAL TECHNOLOGY

Immediate prerequisites of microalgae cultivation system are energy, capital cost, and environmental impact. In most cases, energy and carbon balance must be favorable. Microalgae system requires not only a lot of energy consumption but also a lot of energy for downstream processing (harvesting and extraction). Pond technology seems to have surpassed commercial-scale PBR. However, other parameters related to fluid dynamics still need to be rigorously evaluated. Since the effects of fluid dynamics, such as mixing and turbulence, cause light confinement and light suppression, new design standards that focus on increasing flow rates can significantly increase energy savings by up to 60% (Chiaramonti et al., 2013).

TABLE 1.1
Treatment Efficiency and Microalgae Species

Microalgae Species	Wastewater Type	Treatment Efficiency (%)	Reference
Algal-bacterial symbiosis (*Chlorella þ Nitzchia*)	settled domestic sewage	N: 92 P: 74 COD: 87 BOD: 97	Wang et al. (2010)
Auxenochlorella protothecoides	Concentrated municipal wastewater	TN: 9.8 TP: 13.5 TOC: 16	Renuka et al. (2015)
Chlamydomonas mexicana	Piggery wastewater (filter sterilized)	TN: 3.12 TP: 1.4 TOC:1.45	Renuka et al. (2015)
Chlorella pyrenoidosa	Soybean processing wastewater	TN: 88.8 TP: 70.3 COD: 77.8 NH4-N: 89.1	Renuka et al. (2015)
***Chlamydomonas* sp.**	Industrial wastewater	NH4 +:100 PO4 3-: 33	
Synechocystissalina	Chemical (based products) wastewater collected from the Periyo	NO3–:82.5 NO2–:96.23 PO4 3–:64.52	Emparan et al., (2019)

It is estimated that seawater can help reduce the nutrient and water footprint, thereby promoting the development of algae technology (Yang et al., 2011). Due to fast and dense population growth and urbanization, shortage of freshwater resources is increasing every day. It is estimated that the amount of wastewater generated is 450 billion cubic meters yr^{-1} (Flörke et al., 2013). These wastewaters are being treated in their respective treatment plants for reuse and irrigation (Sato et al., 2013). However, in many developing countries, WWT system faces difficulties in maintenance (Qadir et al., 2010). In the current situation, sewage treatment plants require large amounts of chemicals and operational and maintenance costs, which make the process energy-intensive and expensive (Mo and Zhang, 2012). Microalgae technology used for WWT is very promising and reduces the drawbacks of conventional WWT plants, by providing photosynthetic oxidation for bacterial respiration to eliminate the role of blowers, thereby reducing energy requirements, production of large amounts of algal biomass, and carbon emission reduction etc. (Xin et al., 2016).

1.7 FUTURE RESEARCH PERSPECTIVES AND CONCLUSIONS

Comparing WWT profiles with cultivation strategies can easily identify research gaps. It is worth noting that the presence of rich organic nutrients in wastewater will potentially inhibit the growth of algae. Although many studies using microalgae to treat wastewater have shown better nutrient removal, this algal technology seems to be very promising only in the third stage of wastewater treatment. Identification of bacteria that can recognize each other and participate in wastewater remediation is imminent. Under the influence of light, nitrogen to phosphorus ratio, temperature, pH, and other environmental conditions, more research is needed to establish potential interactions between microalgae strains and bacteria. Rapid growth of strains and the adaptability of microalgae to higher C:N:P ratios will play a key role in the commercialization of cultivation strategies. In addition, other related fields need to be explored, such as the design of PBR, to improve and validate microalgae technology on a commercial level. Since traditional biofilm technology has been proven to be effective, biofilm-based algae WWT is promising and effective on a laboratory scale and requires commercial scale trials. With the goal of developing low-cost flocculants to recover biomass, the scope of further research is imminent. However, conducting more research to determine useful biomass-derived natural products in algae technology can help greatly reduce the investment cost of aquaculture-based products. Using solar energy to dehydrate and dry-harvested biomass may provide another benefit of the viability of microalgae technology on a commercial scale. Overall, wastewater-based algae cultivation systems have many prospects for advancing the technology from transition to commercial scale. WWT's microalgae technology is economical, sustainable, and effective. However, in the current situation, some of the limitations are a setback in bringing the technology to a commercial level. However, each type of microalgae has more specialized characteristics and ability to eradicate various types of pollutants. More studies have shown that new types of microalgae that have not yet been discovered will place great emphasis on finding more effective and stable strains to degrade pollutants. Technological and economic analysis shows that the large-scale implementation of microalgae system for WWT is effective and commercially feasible because biofuels produced by microalgae can compete with conventional fuels. This review focused on current research strategies for subsequent improvements and identifies various limitations of microalgae cultivation techniques for effective establishment and implementation. The flocculation rate, ratio of microalgae to bacteria, is increased, which can greatly reduce pollutants in wastewater. In fact, compared with the related cost-effective harvesting, the application of biofilm technology is expected to remove nutrients compared with suspension systems.

REFERENCES

Abinandan, S., Bhattacharya, R., and Shanthakumar, S. (2015). Efficacy of Chlorella pyrenoidosa and Scenedesmusabundans for nutrient removal in rice mill effluent (paddy-soaked water). *International Journal of Phytoremediation*, *17*(4), 377–381.

Adey, W.H., and Loveland, K. (2011). *Dynamic aquaria: Building living ecosystems*. Elsevier.

Aksu, Z., and Tezer, S. (2005). Biosorption of reactive dyes on the green alga Chlorella Vulgaris. *Process Biochemistry*, *40*(3–4), 1347–1361.

Andrade, C.J., and Andrade, L.M. (2018). Microalgae for bioremediation of textile wastewater. An overview. *MOJ Food Process Technol*, *6*(5), 432–433.

Ansa, E.D.O., Lubberding, H.J., Ampofo, J.A., and Gijzen, H.J. (2011). The role of algae in the removal of Escherichia coli in a tropical eutrophic lake. *Ecological Engineering*, *37*(2), 317–324.

Ansa, E.D.O., Lubberding, H.J., and Gijzen, H.J. (2012). The effect of algal biomass on the removal of faecal coliform from domestic wastewater. *Applied Water Science*, *2*(2), 87–94.

Aslan, S., and Kapdan, I.K. (2006). Batch kinetics of nitrogen and phosphorus removal from synthetic wastewater by algae. *Ecological Engineering*, *28*(1), 64–70.

Azeez, R.A. (2010). A study on the effect of temperature on the treatment of industrial wastewater using chlorella Vulgaris alga. *Algae*, *8*, 9.

Berner, F., Heimann, K., and Sheehan, M. (2015). Microalgal biofilms for biomass production. *Journal of Applied Phycology*, *27*(5), 1793–1804.

Boelee, N.C., Janssen, M., Temmink, H., Shrestha, R., Buisman, C.J.N., and Wijffels, R.H. (2014). Nutrient removal and biomass production in an outdoor pilot-scale phototrophic biofilm reactor for effluent polishing. *Applied Biochemistry and Biotechnology*, *172*(1), 405–422.

Brennan, L., and Owende, P. (2010). Biofuels from microalgae—A review of technologies for production, processing, and extractions of biofuels and co-products. *Renewable and Sustainable Energy Reviews*, *14*(2), 557–577.

Cai, T., Park, S.Y., and Li, Y. (2013). Nutrient recovery from wastewater streams by microalgae: Status and prospects. *Renewable and Sustainable Energy Reviews*, *19*, 360–369.

Chan, A., Salsali, H., and McBean, E. (2014). Heavy metal removal (copper and zinc) in secondary effluent from wastewater treatment plants by microalgae. *ACS Sustainable Chemistry & Engineering*, *2*(2), 130–137.

Chen, X., Li, Z., He, N., Zheng, Y., Li, H., Wang, H., … and Peng, Y. (2018). Nitrogen and phosphorus removal from anaerobically digested wastewater by microalgae cultured in a novel membrane photobioreactor. *Biotechnology for Biofuels*, *11*(1), 1–11.

Chew, K.W., Yap, J.Y., Show, P.L., Suan, N.H., Juan, J.C., Ling, T.C., … and Chang, J.S. (2017). Microalgae biorefinery: High value products perspectives. *Bioresource Technology*, *229*, 53–62.

Chiaramonti, D., Prussi, M., Casini, D., Tredici, M.R., Rodolfi, L., Bassi, N., … and Bondioli, P. (2013). Review of energy balance in raceway ponds for microalgae cultivation: Re-thinking a traditional system is possible. *Applied Energy*, *102*, 101–111.

Choi, H.J. (2015). Biosorption of heavy metals from acid mine drainage by modified sericite and microalgae hybrid system. *Water, Air, & Soil Pollution*, *226*(6), 1–8.

Christenson, L.B., and Sims, R.C. (2012). Rotating algal biofilm reactor and spool harvester for wastewater treatment with biofuels by-products. *Biotechnology and Bioengineering*, *109*(7), 1674–1684.

Christenson, L., and Sims, R. (2011). Production and harvesting of microalgae for wastewater treatment, biofuels, and bioproducts. *Biotechnology Advances*, *29*(6), 686–702.

Chu, W.L., and Phang, S.M. (2019). Biosorption of heavy metals and dyes from industrial effluents by microalgae. In *Microalgae biotechnology for development of biofuel and wastewater treatment* (pp. 599–634). Springer, Singapore.

Craggs, R.J., Adey, W.H., Jenson, K.R., John, M.S.S., Green, F.B., and Oswald, W.J. (1996). Phosphorus removal from wastewater using an algal turf scrubber. *Water Science and Technology*, *33*(7), 191–198.

Cuellar-Bermudez, S.P., Aleman-Nava, G.S., Chandra, R., Garcia-Perez, J.S., Contreras-Angulo, J.R., Markou, G., …, and Parra-Saldivar, R. (2017). Nutrients utilization and contaminants removal. A review of two approaches of algae and cyanobacteria in wastewater. *Algal Research*, *24*, 438–449.

D'Aiuto, P.E., Patt, J.M., Albano, J.P., Shatters, R.G., and Evens, T.J. (2015). Algal turf scrubbers: Periphyton production and nutrient recovery on a South Florida citrus farm. *Ecological Engineering*, *75*, 404–412.

Dar, R.A., Sharma, N., Kaur, K., and Phutela, U.G. (2019). Feasibility of microalgal technologies in pathogen removal from wastewater. In *Application of microalgae in wastewater treatment* (pp. 237–268). Springer, Cham.

Das, A., Adhikari, S., and Kundu, P. (2019). Bioremediation of wastewater using microalgae. In *Environmental biotechnology for soil and wastewater implications on ecosystems* (pp. 55–60). Springer, Singapore.

Delattre, C., Pierre, G., Laroche, C., and Michaud, P. (2016). Production, extraction and characterization of microalgal and cyanobacterial exopolysaccharides. *Biotechnology Advances, 34*(7), 1159–1179.

Diniz, G.S., Silva, A.F., Araújo, O.Q., and Chaloub, R.M. (2017). The potential of microalgal biomass production for biotechnological purposes using wastewater resources. *Journal of Applied Phycology, 29*(2), 821–832.

El-Sheekh, M.M., Farghl, A.A., Galal, H.R., and Bayoumi, H.S. (2016). Bioremediation of different types of polluted water using microalgae. *RendicontiLincei, 27*(2), 401–410.

Emparan, Q., Harun, R., and Danquah, M.K. (2019). Role of phycoremediation for nutrient removal from wastewaters: A review. *Applied Ecology and Environmental Research, 17*, 889–915.

Farooq, W., Lee, Y.C., Ryu, B.G., Kim, B.H., Kim, H.S., Choi, Y.E., and Yang, J.W. (2013). Two-stage cultivation of two Chlorella sp. strains by simultaneous treatment of brewery wastewater and maximizing lipid productivity. *Bioresource Technology, 132*, 230–238.

Fitch, M.W., Pearson, N., Richards, G., and Burken, J.G. (1998). Biological fixed-film systems. *Water Environment Research, 70*(4), 495–518.

Flörke, M., Kynast, E., Bärlund, I., Eisner, S., Wimmer, F., and Alcamo, J. (2013). Domestic and industrial water uses of the past 60 years as a mirror of socio-economic development: A global simulation study. *Global Environmental Change, 23*(1), 144–156.

Gajda, I., Greenman, J., Melhuish, C., and Ieropoulos, I. (2015). Self-sustainable electricity production from algae grown in a microbial fuel cell system. *Biomass and Bioenergy, 82*, 87–93.

Gupta, P.L., Lee, S.M., and Choi, H.J. (2015). A mini review: Photobioreactors for large scale algal cultivation. *World Journal of Microbiology and Biotechnology, 31*(9), 1409–1417.

Gutzeit, G., Lorch, D., Weber, A., Engels, M., and Neis, U. (2005). Bioflocculent algal–bacterial biomass improves low-cost wastewater treatment. *Water Science and Technology, 52*(12), 9–18.

Han, F., Huang, J., Li, Y., Wang, W., Wang, J., Fan, J., and Shen, G. (2012). Enhancement of microalgal biomass and lipid productivities by a model of photoautotrophic culture with heterotrophic cells as seed. *Bioresource Technology, 118*, 431–437.

Han, S.F., Jin, W.B., Tu, R.J., and Wu, W.M. (2015). Biofuel production from microalgae as feedstock: Current status and potential. *Critical Reviews in Biotechnology, 35*(2), 255–268.

He, J., and Chen, J.P. (2014). A comprehensive review on biosorption of heavy metals by algal biomass: Materials, performances, chemistry, and modeling simulation tools. *Bioresource Technology, 160*, 67–78.

Hussein, M.H., Abdullah, A.M., Eladal, E.G., and El-Din, N.B. (2016). Phycoremediation of some pesticides by microchlorophyte alga, Chlorella Sp. *Journal of Fertilizers & Pesticides, 7*(173), 2.

Jacinto, M.L.J., David, C.P.C., Perez, T.R., and De Jesus, B.R. (2009). Comparative efficiency of algal biofilters in the removal of chromium and copper from wastewater. *Ecological Engineering, 35*(5), 856–860.

Ji, M.K., Yun, H.S., Hwang, B.S., Kabra, A.N., Jeon, B.H., and Choi, J. (2016). Mixotrophic cultivation of Nephroselmis sp. using industrial wastewater for enhanced microalgal biomass production. *Ecological Engineering, 95*, 527–533.

Klinthong, W., Yang, Y.H., Huang, C.H., and Tan, C.S. (2015). A review: Microalgae and their applications in CO_2 capture and renewable energy. *Aerosol and Air Quality Research, 15*(2), 712–742.

Kong, Q.X., Li, L., Martinez, B., Chen, P., and Ruan, R. (2010). Culture of microalgae Chlamydomonasreinhardtii in wastewater for biomass feedstock production. *Applied Biochemistry and Biotechnology*, *160*(1), 9–18.

Koppel, D.J., Adams, M.S., King, C.K., and Jolley, D.F. (2019). Preliminary study of cellular metal accumulation in two Antarctic marine microalgae–implications for mixture interactivity and dietary risk. *Environmental Pollution*, *252*, 1582–1592.

Kumar, S.D., Santhanam, P., Nandakumar, R., Anath, S., Prasath, B.B., Devi, A.S., … and Ananthi, P. (2014). Preliminary study on the dye removal efficacy of immobilized marine and freshwater microalgal beads from textile wastewater. *African Journal of Biotechnology*, *13*(22), 2288–2294.

Lee, C.S., Oh, H.S., Oh, H.M., Kim, H.S., and Ahn, C.Y. (2016). Two-phase photoperiodic cultivation of algal–bacterial consortia for high biomass production and efficient nutrient removal from municipal wastewater. *Bioresource Technology*, *200*, 867–875.

Lee, J., Lee, J., Shukla, S.K., Park, J., and Lee, T.K. (2016). Effect of algal inoculation on COD and nitrogen removal, and indigenous bacterial dynamics in municipal wastewater. *Journal of Microbiology and Biotechnology*, *26*(5), 900–908.

Lee, K., and Lee, C.G. (2001). Effect of light/dark cycles on wastewater treatments by microalgae. *Biotechnology and Bioprocess Engineering*, *6*(3), 194–199.

Liang, Z., Liu, Y., Ge, F., Xu, Y., Tao, N., Peng, F., and Wong, M. (2013). Efficiency assessment and pH effect in removing nitrogen and phosphorus by algae-bacteria combined system of Chlorella Vulgaris and Bacillus licheniformis. *Chemosphere*, *92*(10), 1383–1389.

Liu, L., Hall, G., and Champagne, P. (2020). The role of algae in the removal and inactivation of pathogenic indicator organisms in wastewater stabilization pond systems. *Algal Research*, *46*, 101777.

Luo, S., Berges, J.A., He, Z., and Young, E.B. (2017). Algal-microbial community collaboration for energy recovery and nutrient remediation from wastewater in integrated photobioelectrochemical systems. *Algal Research*, *24*, 527–539.

Ma, X., Zheng, H., Addy, M., Anderson, E., Liu, Y., Chen, P., and Ruan, R. (2016). Cultivation of Chlorella Vulgaris in wastewater with waste glycerol: Strategies for improving nutrients removal and enhancing lipid production. *Bioresource Technology*, *207*, 252–261.

Mehrabadi, A., Craggs, R., and Farid, M.M. (2015). Wastewater treatment high rate algal ponds (WWT HRAP) for low-cost biofuel production. *Bioresource Technology*, *184*, 202–214.

Miao, M.S., Yao, X.D., Shu, L., Yan, Y.J., Wang, Z., Li, N., … and Kong, Q. (2016). Mixotrophic growth and biochemical analysis of Chlorella Vulgaris cultivated with synthetic domestic wastewater. *International Biodeterioration & Biodegradation*, *113*, 120–125.

Milledge, J.J., and Heaven, S. (2013). A review of the harvesting of micro-algae for biofuel production. *Reviews in Environmental Science and Bio/Technology*, *12*(2), 165–178.

Min, M., Wang, L., Li, Y., Mohr, M.J., Hu, B., Zhou, W., … and Ruan, R. (2011). Cultivating Chlorella sp. in a pilot-scale photobioreactor using centrate wastewater for microalgae biomass production and wastewater nutrient removal. *Applied Biochemistry and Biotechnology*, *165*(1), 123–137.

Mo, W., and Zhang, Q. (2012). Can municipal wastewater treatment systems be carbon neutral? *Journal of Environmental Management*, *112*, 360–367.

Mulbry, W., Kondrad, S., Pizarro, C., and Kebede-Westhead, E. (2008). Treatment of dairy manure effluent using freshwater algae: Algal productivity and recovery of manure nutrients using pilot-scale algal turf scrubbers. *Bioresource Technology*, *99*(17), 8137–8142.

Onuoha, O. (2012). *Combined Global and African Ranking - 25 country populations with the least sustainable access to improved/clean water sources*. Africa Public Health Info.

Osundeko, O., and Pittman, J.K. (2014). Implications of sludge liquor addition for wastewater-based open pond cultivation of microalgae for biofuel generation and pollutant remediation. *Bioresource Technology*, *152*, 355–363.

Oswald, J.A. (1988). Large-scale algal culture systems (engineering aspects). *Micro-algal Biotechnology*. In Borowitzka, L.J., and Borowitzka, M.A. (Eds.), pp. 357–394. Cambridge University Press, Cambridge.

Park, J.B.K., and Craggs, R.J. (2011). Algal production in wastewater treatment high rate algal ponds for potential biofuel use. *Water Science and Technology*, *63*(10), 2403–2410.

Pizarro, C., Mulbry, W., Blersch, D., and Kangas, P. (2006). An economic assessment of algal turf scrubber technology for treatment of dairy manure effluent. *Ecological Engineering*, *26*(4), 321–327.

Posadas, E., Muñoz, A., García-González, M.C., Muñoz, R., and García-Encina, P.A. (2015). A case study of a pilot high rate algal pond for the treatment of fish farm and domestic wastewaters. *Journal of Chemical Technology & Biotechnology*, *90*(6), 1094–1101.

Qadir, M., Wichelns, D., Raschid-Sally, L., McCornick, P.G., Drechsel, P., Bahri, A., and Minhas, P.S. (2010). The challenges of wastewater irrigation in developing countries. *Agricultural Water Management*, *97*(4), 561–568.

Qureshi, N., Annous, B.A., Ezeji, T.C., Karcher, P., and Maddox, I.S. (2005). Biofilm reactors for industrial bioconversion processes: Employing potential of enhanced reaction rates. *Microbial Cell Factories*, *4*(1), 1–21.

Ramakrishnan, B., Megharaj, M., Venkateswarlu, K., Naidu, R., and Sethunathan, N. (2010). The impacts of environmental pollutants on microalgae and cyanobacteria. *Critical Reviews in Environmental Science and Technology*, *40*(8), 699–821.

Rasoul-Amini, S., Montazeri-Najafabady, N., Shaker, S., Safari, A., Kazemi, A., Mousavi, P., and Ghasemi, Y. (2014). Removal of nitrogen and phosphorus from wastewater using microalgae free cells in bath culture system. *Biocatalysis and Agricultural Biotechnology*, *3*(2), 126–131.

Rathod, H. (2014, August). Algae based wastewater treatment. In *A Seminar Report of Master of Technology in Civil Engineering*. Roorkee, Uttarakhand, India.

Renuka, N., Sood, A., Prasanna, R., and Ahluwalia, A.S. (2015). Phycoremediation of wastewaters: A synergistic approach using microalgae for bioremediation and biomass generation. *International Journal of Environmental Science and Technology*, *12*(4), 1443–1460.

Salama, E.S., Kurade, M.B., Abou-Shanab, R.A., El-Dalatony, M.M., Yang, I.S., Min, B., and Jeon, B.H. (2017). Recent progress in microalgal biomass production coupled with wastewater treatment for biofuel generation. *Renewable and Sustainable Energy Reviews*, *79*, 1189–1211.

Salama, E.S., Roh, H.S., Dev, S., Khan, M.A., Abou-Shanab, R.A., Chang, S.W., and Jeon, B.H. (2019). Algae as a green technology for heavy metals removal from various wastewater. *World Journal of Microbiology and Biotechnology*, *35*(5), 1–19.

Sato, T., Qadir, M., Yamamoto, S., Endo, T., and Zahoor, A. (2013). Global, regional, and country level need for data on wastewater generation, treatment, and use. *Agricultural Water Management*, *130*, 1–13.

Schnurr, P.J., and Allen, D.G. (2015). Factors affecting algae biofilm growth and lipid production: A review. *Renewable and Sustainable Energy Reviews*, *52*, 418–429.

Schnurr, P.J., Espie, G.S., and Allen, D.G. (2013). Algae biofilm growth and the potential to stimulate lipid accumulation through nutrient starvation. *Bioresource Technology*, *136*, 337–344.

Singh, R., Paul, D., and Jain, R.K. (2006). Biofilms: Implications in bioremediation. *Trends in Microbiology*, *14*, 389–397.

Sukačová, K., Trtílek, M., and Rataj, T. (2015). Phosphorus removal using a microalgal biofilm in a new biofilm photobioreactor for tertiary wastewater treatment. *Water Research*, *71*, 55–63.

Torres, E.M., Hess, D., McNeil, B.T., Guy, T., and Quinn, J.C. (2017). Impact of inorganic contaminants on microalgae productivity and bioremediation potential. *Ecotoxicology and Environmental Safety*, *139*, 367–376.

Wang, H., Xiong, H., Hui, Z., and Zeng, X. (2012). Mixotrophic cultivation of Chlorella pyrenoidosa with diluted primary piggery wastewater to produce lipids. *Bioresource Technology*, *104*, 215–220.

Wang, L., Min, M., Li, Y., Chen, P., Chen, Y., Liu, Y., ... and Ruan, R. (2010). Cultivation of green algae Chlorella sp. in different wastewaters from municipal wastewater treatment plant. *Applied Biochemistry and Biotechnology*, *162*(4), 1174–1186.

White, H.K., Reimers, C.E., Cordes, E.E., Dilly, G.F., and Girguis, P.R. (2009). Quantitative population dynamics of microbial communities in plankton-fed microbial fuel cells. *The ISME Journal*, *3*(6), 635–646.

Woertz, I., Feffer, A., Lundquist, T., et al. (2009). Algae grown on dairy and municipal wastewater for simultaneous nutrient removal and lipid production for biofuel feedstock. *Journal of Environmental Engineering*, *135*, 1115–1122.

Xiao, R., and Zheng, Y. (2016). Overview of microalgal extracellular polymeric substances (EPS) and their applications. *Biotechnology Advances*, *34*(7), 1225–1244.

Xin, C., Addy, M.M., Zhao, J., Cheng, Y., Cheng, S., Mu, D., ... and Ruan, R. (2016). Comprehensive techno-economic analysis of wastewater-based algal biofuel production: A case study. *Bioresource Technology*, *211*, 584–593.

Xiong, J.Q., Kurade, M.B., Abou-Shanab, R.A., Ji, M.K., Choi, J., Kim, J.O., and Jeon, B.H. (2016). Biodegradation of carbamazepine using freshwater microalgae Chlamydomonasmexicana and Scenedesmusobliquus and the determination of its metabolic fate. *Bioresource Technology*, *205*, 183–190.

Yang, J., Xu, M., Zhang, X., Hu, Q., Sommerfeld, M., and Chen, Y. (2011). Life-cycle analysis on biodiesel production from microalgae: Water footprint and nutrients balance. *Bioresource Technology*, *102*(1), 159–165.

Yu, K.L., Show, P.L., Ong, H.C., Ling, T.C., Chen, W.H., and Salleh, M.A.M. (2018). Biochar production from microalgae cultivation through pyrolysis as a sustainable carbon sequestration and biorefinery approach. *Clean Technologies and Environmental Policy*, *20*(9), 2047–2055.

Zhang, Y., Su, H., Zhong, Y., Zhang, C., Shen, Z., Sang, W., ..., and Zhou, X. (2012). The effect of bacterial contamination on the heterotrophic cultivation of Chlorella pyrenoidosa in wastewater from the production of soybean products. *Water Research*, *46*(17), 5509–5516.

2 Recent Progress of Phytoremediation-Based Technologies for Industrial Wastewater Treatment

Rustiana Yuliasni
Center of Industrial Pollution Prevention Technology, The Ministry of Industry Republic of Indonesia, Semarang, Central Java, Indonesia

Setyo Budi Kurniawan
Department of Chemical and Process Engineering, Faculty of Engineering and Built Environment, National University of Malaysia (UKM), Bangi, Selangor, Malaysia

Bekti Marlena
Center of Industrial Pollution Prevention Technology, The Ministry of Industry Republic of Indonesia, Semarang, Central Java, Indonesia

Mohamad Rusdi Hidayat
Institute for Industrial Research and Standarization of Pontianak, The Ministry of Industry Republic of Indonesia, Pontianak, West Kalimantan, Indonesia

Abudukeremu Kadier and Peng Cheng Ma
Laboratory of Environmental Science and Technology, The Xinjiang Technical Institute of Physics and Chemistry, Key Laboratory of Functional Materials and Devices for Special Environments, Chinese Academy of Sciences, Urumqi, China

CONTENTS

DOI: 10.1201/9781003165101-2

2.1 INTRODUCTION

Conventional wastewater treatments are not completely effective methods for water-contaminant removal. A trace concentration of toxic contaminants can still be found in wastewater effluents. Thus, an alternative technology to reduce the contaminant concentration to a safe level is necessary. Different types of wastewater treatment technology are introduced. However, most of these technologies are considered to be high-energy requirements: high carbon emission, excess sludge discharge, and high maintenance cost (Mustafa and Hayder, 2020). A sustainable management of aquatic ecosystem needs eco-friendly and low-cost remediation methods. Aquatic plants have the potential to remove inorganic and organic pollutants. Phytoremediation is defined as a bioremediation that uses plants for wastewater remediation, using the plant's roof to adsorb nutrients in the wastewater. Specific species of plants even have the ability to accumulate certain pollutants. Phytoremediation has been proven to be more efficient, cost effective, and more environment friendly than conventional treatment.

There are plants that have high phytoremediation ability, such as *Brassica juncea*, *Arundo donax* L. *Miscanthus* sp, *Typha latifolia,* and *Thelypteris palustris* for heavy metals removal such as Zn and Cu, by using bioaccumulation mechanism (Ullah et al., 2015). *Salvinia molesta* and *Pistia stratiotes* also have been widely used for the treatment of agricultural, domestic, and industrial wastewater (Mustafa and Hayder, 2020). Type of plants are not the only main factors for successful phytoremediation process; the role of rhizosphere-associated microorganisms is also important. Microorganisms help to improve the phytoremediation process through biosorption and bio-augmentation. Organisms such as *Acidovorax, Alcaligenes, Bacillus 95 mycobacterium, Paenibacillus, Pseudomonas*, and *Rhodococcus* have been reported to enhance the phytoremediation process (Sharma et al., 2021).

However, phytoremediation of polluted water in wetland-type reactors has been mostly studied as black box. The method to measure the performance is only based on pollutant-removal efficiency, and only very limited information is available concerning the pollutant-removal mechanisms and process dynamics in these systems. This chapter briefly reviews basic processes of phytoremediation, its mechanisms and parameters, and its interaction between rhizo-remediation and microbe-plant. In addition, it also elaborates on phytoremediation challenges and strategies for full-scale application, its techniques to remove both organic and inorganic contaminants by aquatic plants in water, and some example of applications in industries.

2.2 PRINCIPLES OF PHYTOREMEDIATION

The concept of phytoremediation must be differentiated from bioremediation. The bioremediation process is merely assisted by heterotrophic bacteria that is responsible for organic contaminant degradation and mineralization, as well as metals and other element accumulation and inorganic compounds oxidation (McCutcheon and Jørgensen, 2018). Alternatively, the phytoremediation process is based on the role of photoautotroph bacteria to treat contaminants via mechanisms, such as:

a. release organic matter as their metabolism products (during growth and maintenance), thus improving the number of heterotrophs bacteria.
b. pump the oxygen into the plant-root zone and also deposit secondary metabolites during root die-back in the rhizosphere to boost the number of aerobic, facultative, or anaerobic organisms to degrade or accumulate contaminants.
c. transport pollutants into active microbial zones by evapotranspiration, blockage of flows, or other means.
d. in more details, phytoremediation mechanisms can break down into types, namely: phytodegradation, phytoextraction, phytovolatilization, phytofiltration, and phytostabilization, as shown in Table 2.1.

2.3 INTERACTION BETWEEN RHIZO-REMEDIATION AND MICROBE-PLANT IN PHYTOREMEDIATION

Rhizosphere is the most important area during phytoremediation (Purwanti et al., 2020). Rhizosphere is the place where pollutants have contact with the treatment agent (plant) (Al-Ajalin et al., 2020a; Ismail et al., 2020). Plants root played an important role in the removal of pollutant from wastewater (Al-Ajalin et al., 2020b). Beside the root, there are also microbes (known as rhizobacteria) that also greatly support the degradation of pollutants in the rhizophore (Jehawi et al., 2020). Plant roots and microbes perform interactions that lead to the removal of contaminants from the contaminated medium, as illustrated on Figure 2.1.

Four major interactions in rhizosphere occur during the phytoremediation of pollutants from wastewater: phytostimulation (Hawrot-Paw et al., 2019), rhizofiltration (Rahman and Hasegawa, 2011), rhizodegradation (Imron et al., 2019b; Kadir et al., 2020), and phytostabilization (Bolan et al., 2011). Phytostimulation is a process in which plants release their exudates in the rhizosphere (Backer et al., 2018). Released exudates near the root area provide a good environment for rhizobacteria to grow optimally (Abdullah et al., 2020). The release of exudates stimulates the growth of rhizobacteria, which performs symbiotic interactions (Shahid et al., 2020). The phytostimulation cannot be separated from rhizodegradation. Rhizodegradation is the mechanisms where rhizobacteria perform the degradation of pollutants in rhizosphere (Almansoory et al., 2020; Imron et al., 2020). The better the growth of rhizobacteria, the more degradation of pollutants will be obtained. Rhizodegradation mostly occurred during the treatment of organic materials-rich wastewater (Tangahu et al., 2019).

TABLE 2.1
Type of Phytoremediation Mechanisms, Their Affecting Factors and Applications

Type of Phytoremediation	Mechanisms	Affecting Factors	Applications	Reference
Phytodegradation/ phyto-oxidation	Phytodegradation occurs when aquatic and terrestrial plants take up, store, and biochemically degrade or transform organic compounds to harmless by-products, products used to create new plant biomass, or by-products that are further broken down by microbes and other processes to less-harmful compounds. Growth and senescence enzymes, sometimes in series, are involved in plant metabolism or detoxification of reductive and different parts of the plant.	Concentration and composition of pollutant, plant species, and soil conditions.	Soil, sediment sludges, groundwater and surface water, wetlands, wastewaters, and air contaminated with compounds	(Kagalkar et al., 2011; Park et al., 2011)
Phytoextraction	Contaminants is transferred to harvestable plant tissues by hyperaccumulation	Contaminant concentration, the depth of the contamination in the soil, the possibility of leach of pollutants into groundwater	Soil	(Wang et al., 2020)
Phytovolatilisation	Volatilisation by leaves. Transformation of toxic substances into less toxic.	The possibility of re-deposition of pollutant back into ecosystem by precipitation (elemental	Soils, sediments, sludges, wetlands, and groundwater up	(Epa, 2019)
Phytofiltration	Accumulation of contaminants in rhizosphere	The plant must have high metal-resistant, high adsorption surface, high tolerance of hypoxia. Long-term maintenance depends on type of contaminant and depth, hinders plant growth, highly species specifi	Wetlands, wastewater, landfill leachates, and groundwater contaminate with metals, radionuclides, organic chemicals, nitrate, ammonium, phosphate, and pathogens.	(Sandhi et al., 2018)

Phytostabilization	Revegetation to prevent erosion and sorbed pollutant transport	Plants control pH, soil gases, and redox that cause speciation, precipitation, and sorption to form stable mineral deposits (effects ecosystem succession unknown on long-term stability and thus sustainability)	Soil, mine tailings, wetlands, and leachate pond sediments contaminated with metals, phenols, anilines, and some pesticides

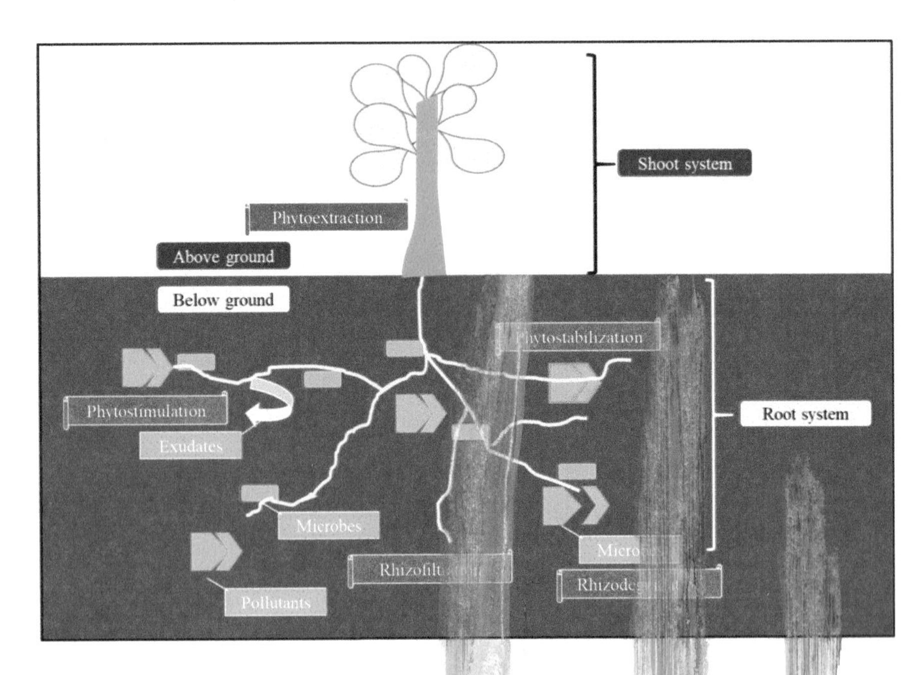

FIGURE 2.1 Microbe-plants interaction during phytoremediation.

For heavy metals containing wastewater, rhizobacteria may act as a stabilization agent that transforms the ionic state metals into stable state (Imron et al., 2019a; Kurniawan et al., 2018; Titah et al., 2018). Rhizobacteria may also perform bioaccumulation, which then leads to the stabilization of heavy metals inside cells (Purwanti et al., 2019; Titah et al., 2019). In addition to the rhizobacterial processing heavy metals, plant exudates contain complex compounds that may increase the solubility of metals (to be treated further by rhizobacteria) or to bind directly with heavy metals to produce complex metal-exudates that then stabilize in the rhizosphere (phytostabilization) (Dakora and Phillips, 2002). Plant roots also perform the physical treatment of wastewater by performing the screening of bulk compounds in their roots. This mechanism mostly occurred in the treatment of pollutants using fibrous root-type species (Elias et al., 2014). After performing several mechanisms in rhizosphere, plants then perform phytoextraction in which they absorb pollutants via a transfer mechanism to bioconcentrate it into their cell (Purwanti et al., 2020). Phytoextraction can occur directly to the pollutant and also its intermediate compounds (after being degraded by rhizobacteria). There are no significant differences in mechanisms occurring between the treatment of wastewater using sub-surface or free-surface constructed wetlands. The major differences are the species used and the contaminated medium that needs to be treated (Kadir et al., 2020; Purwanti et al., 2018b, 2018a).

2.4 RHIZO-REMEDIATION AND MICROBE-PLANT INTERACTION IN PHYTOREMEDIATION

Despite the many advantages of phytoremediation application for industrial wastewater treatment, this method still has some challenges to face during application.

Challenges

- Environmental condition
- Contact with pollutant is limited in the rhizosphere
- Capability of plant to withstand polutant
- Area requirement
- Time requirement
- Biomass handling

Strategies

- Controlled environment (under optimum condition)
- Design of appropriate constructed wetland system
- Range finding/phytotoxicity test
- Combined treatment system
- Convertion of biomass into valuable products

FIGURE 2.2 Challenges and strategies of phytoremediation application.

Some challenges of phytoremediation application and strategies that may cover the challenges are summarized in Figure 2.2.

Phytoremediation needs certain conditions to work well, including the requirement of sunlight (Miranda et al., 2020), specific nutrients for plant growth (Bansal et al., 2019; Varma et al., 2021), temperature (Mao et al., 2015), humidity (Armstrong et al., 1992), etc. These requirements need to be fulfilled during application to obtain the best removal performance. Phytoremediation is considered to be very suitable for use in the tropical countries (Ahmad et al., 2017) due to the availability of sunlight throughout the year and optimum temperature and humidity for plant growth, while in subtropical countries, a controlled environment is essential (Ismail et al., 2019). Greenhouse treatment is suggested to be applied to maintain the optimum environmental condition for plants to treat pollutants. Under controlled environments, plants will be able to maintain their performance throughout the year, which may lead to the desired removal efficiencies.

Rhizosphere is the most important area in phytoremediation since the contact of pollutants and treatment agents occurs there (Kamaruzzaman et al., 2019). This may become a challenge when plants root lack good contact with pollutants. To overcome this issue, the design of appropriate constructed wetland needs to be conducted before the application (Purwanti et al., 2018b). In designing appropriate constructed wetlands, several major criteria need to be considered: characteristic of pollutants (Mostafa, 2015; Sharuddin et al., 2018), fate of pollutants (Logeshwaran et al., 2018), type of wetland to be used (Al-Ajalin et al., 2020c), type of plants, plant growth (related to the root growth and penetration in medium) (Schwammberger et al., 2019), medium for plants (Sun et al., 2007), and depth of constructed wetland (Al-Ajalin et al., 2020a).

By using biological methods, researchers need to aware that certain concentrations of pollutants may disturb the performance of phytoremediation. Only certain plants that can survive high loading of pollutants may perform a better removal in treatment (Kwoczynski and Čmelík, 2021). A limited pollutant concentration can be tolerated by plants (Abdullah et al., 2020). To avoid the death of plants, which may lead to decreased removal performance, the selection of plant species and range finding/phytotoxicity tests need to be conducted before the application of phytoremediation (Purwanti et al., 2018a; 2019). Several criteria in the selection of plant species includes their capability to withstand high pollutant loads (Abdullah et al., 2020), the removal performance (Kurniawan et al., 2020), and capability to growth (perennial plants are more preferable) (Al-Baldawi et al., 2015). The range finding/phytotoxicity test needs to be performed to obtain the maximum concentration of pollutant that can be treated by the utilized species. If plants can withstand 100% concentration of pollutant, then the plants can be used as the primary treatment technology. Secondary or tertiary treatment options should be chosen if plants can withstand lower concentrations of pollutants.

Application of phytoremediation to treat industrial wastewater requires large area; it is also considered time consuming (Abdullah et al., 2020). These issues are closely related to the rate of pollutants' degradation by plants during treatment. Biological treatment has different reactions as compared to chemical treatment (Imron et al., 2020). In chemical treatment, the stoichiometry of reaction controls the degradation of pollutants based on the equilibrium of reactants and products (Kis et al., 2017). In biological treatment, the capability of plants cannot be simply calculated as reactants and products equilibrium due to the complex mechanisms that involve many factors occurring during treatment (Karpowicz et al., 2020; Nottingham et al., 2018). To overcome these issues, most researchers suggest the utilization of the phytoremediation technique as a secondary or tertiary treatment to cleanse wastewater before discharge into water bodies. Chemical treatment is suggested as a primary treatment, which may reduce the pollutant load in phytoremediation stage that may produce better removal rate, reducing the required time and surface area for treatment.

As plants grow during the treatment, plant biomass is produced; the amount can be considered as abundant. If phytoremediation were applied to treat toxic substances (commonly heavy metals), produced plant biomass need to be handled following the standard procedure of handling toxic substances (Kwoczynski and Čmelík, 2021). If phytoremediation were applied to treat organic-rich or nutrient-rich wastewater, several conversion possibilities are available. Several biomass utilization studies have been successfully applied to convert biomass into animal feed (Kadir et al., 2020), biochar (Das et al., 2021), adsorbent (Alshekhli et al., 2020), biofuel (Correa et al., 2019; Rezania et al., 2020), and even fertilizer (Diacono et al., 2019; Kurniawan et al., 2020). With these conversion options, wastewater treatment using phytoremediation may lead to the cleaner production strategy by using a by-product treatment.

2.5 PHYTOREMEDIATION TECHNIQUES BY AQUATIC PLANTS FOR BOTH ORGANIC AND INORGANIC CONTAMINANTS REMOVAL IN WATER

2.5.1 Aquatic Plants Selection

Aquatic plants are required in phytoremediation for degrading and removing contaminants within aquatic environments. These plants include ferns, pteridophytes, and freshwater-adapted angiosperms. Aquatic plants are preferable to terrestrial plants for wastewater treatment because of their faster growth rate, larger biomass production, and better contaminant-removal ability due to direct contact with the wastewater. The effectiveness of these plants in phytoremediation can be assessed by estimating the contaminants removed from the target area. Not only for remediation purposes, many such aquatic plants also serve as bioindicators and biomonitors.

In addition, some key principles that need to be considered in operating a phytoremediation system are as follows: a) identifying the suitable and efficient aquatic plants for the phytoremediation system; b) uptake of dissolved nutrients (e.g. N, P, and metals) by the aquatic plants; and c) harvesting process and utilization of the plant biomass generated from the phytoremediation system. Regular harvest of the aquatic plant biomass from a remediation site is necessary. Otherwise, the plants' biomass will be decomposed and subsequently release the stored contaminants back to the aquatic environment (Kumwimba et al., 2020).

Selection of aquatic plants that can grow well while degrading targeted contaminants is critical. Some plants commonly used for phytoremediation could experience disrupted growth if exposed to a high level of contaminants. The toxicity effects of the contaminants against aquatic plants are varied. Some negative responses of aquatic plants toward aquatic contaminants are growth reduction, wilting, chlorosis, reduction of roots and shoots length or volume, chlorophyll reduction, reduction in photosynthetic activity, and plant mortality. For instance, exposure of high levels of cadmium and zinc to water hyacinth (*Eichornia crassipes*) resulted in reduced growth, as determined from biomass production, survival rate, and crown root number. Another study by de Campos et al. (2019) that exposed water lettuce (*Pistia stratiotes*) with a high level of arsenite showed that although *P. stratiotes* could maintain its biomass, there had been a significant reduction in the root volume, chlorosis in the leaves, and damage in the cell membranes.

The ability of aquatic plants to reduce contaminants varies between plants. Therefore, to reduce the unfavorable effects on the plants' growth in a phytoremediation system, it is necessary to pay attention to the characteristics of the selected plants. The ideal characteristics of aquatic plants used for phytoremediators are as follows: high growth rate, production of more above-ground biomass, widely distributed and highly branched root system, high bioaccumulation potential, ability to transform or degrade contaminants, ability to regulate chemical speciation, capacity to treat both organic and inorganic contaminants, a greater accumulation of the target heavy metals from soil (bioconcentration factor>1), translocation of the accumulated heavy metals from roots to shoots (translocation factor>1), tolerance to the toxic effects of the target heavy metals, good adaptation to prevailing environmental and climatic conditions, resistance to pathogens and pests, easy cultivation and harvest, and repulsion to herbivores to avoid food-chain contamination.

Another primary factor that needs to be considered in the utilization of aquatic plants in a phytoremediation system is understanding the characteristics of the wastewater to be treated. Wastewater is a mixture of pure water with a large number of chemicals (including organic and inorganic chemicals) and heavy metals produced from domestic, agriculture, industrial, and commercial activities. Organic contaminants can be categorized into persistent organic pollutants (POP)/xenobiotics (i.e. dioxins, polycyclic aromatic hydrocarbons, and polychlorinated biphenyls), pesticides (i.e. glyphosate, hexachlorocyclohexane, fenhexamid, and deltamethrin), and pharmaceutical and personal-care products (PPCPs) (i.e. antibiotics, hormones, and pain relief medication). Meanwhile, primary inorganic contaminants are nutrients (i.e., N, P, and K) and metalloid elements (i.e., Fe, Al, Pb, Ni, Cd, and Cu). The existence of these various pollutants in the environment needs serious attention since they can cause various harmful effects. Potential adverse effects of those contaminants on the surrounding environment and living things are as follows: eutrophication, chronic toxicity, endocrine disruption, and antibiotic resistance.

2.5.2 Types of Aquatic Plants

Aquatic plants have earned an immense reputation due to their capacity to clean up contaminated water bodies. With their extensive roots system, these plants become the best option for degrading contaminants in a phytoremediation system. Based on their growth form, aquatic plants can be classified into free-floating, submerged, and emergent plants.

Free-floating aquatic plants are plants with floating leaves and submerged roots. Several free-floating aquatic plants have been studied extensively and approved to be applied in different phytoremediation systems. Some recognized free-floating aquatic plants are duckweeds (*Lemna*, *Spirodela*, and *Wolffia*), water hyacinth (*Eichhornia*), water ferns (*Salvinia*, *Azolla*), and water lettuce (*Pistia*). Those plants are known for having the capability to clean up a wide variety of inorganic and organic contaminants, heavy metals, pesticides, and nutrients from various sources, such as industrial and domestic wastewater, sewage, and agricultural runoff. Moreover, those plants can grow in polluted sites with tremendous variation in temperature, pH, and nutrient level.

Submerged aquatic plants are plants that usually grow underwater and are rooted in mud. Their leaves are the main part for contaminants uptake. Some famous submerged plants that have been studied are watermilfoil (*Myriophyllum*), coontail or hornwort (*Ceratophyllum demersum*), pondweed (*Potamogeton*), Esthwaite waterweed (*Hydrilla*), and water mint (*Mentha aquatica*). Most of these plants are commonly found in slow-moving streams, ponds, and lakes. Additionally, the effectiveness of these plants in removing contaminants depends on different factors, such as contaminant types and their concentration, pH, and temperature (Dhir, 2013 and Javed et al., 2019).

Emergent aquatic plants are plants usually found on submerged soil where the water table is 0.5 m below the soil. These plants grow their shoots and leaves above the water, while keeping their roots beneath the surface. Cattails (*Typha*), bulrush (*Scirpus*), common reed (*Phragmites australis*), reed canary grass (*Phalaris arundinacea*), and foxtail flatsedge (*Cyperus alopecuroides*) are well-known emergent aquatic plants that can effectively be used for phytoremediation (Ali et al., 2020). Emergent plant species have received considerable attention in nutrient phytoremediation and are often deployed in constructed wetlands because they are easier to harvest.

Furthermore, another type of plant that becomes a new interest in phytoremediation is the transgenic plant. Transgenic plants were engineered so that specific genes in the plants can increase their metabolism and enhance the detoxification process of organic pollutants for more effective phytoremediation. In this approach, incorporated genes secrete enzymes, which degrade organic pollutants in the rhizospheric zone. This might solve the problem in plant harvesting and handling loaded with toxic metals, as all the metal detoxification and removal process occur in the rhizosphere by roots. Engineered *Arabidopsis thaliana* and *Nicotiana tabaccum* are examples of transgenic plants effective for removing heavy metals, cadmium, and mercury.

Different species of aquatic plants have been long studied for their potential in phytoremediation with notable successes. Table 2.2 presents some common aquatic plants used in phytoremediation studies in recent years. However, it should be noted that contaminants' degradation efficiencies depend on various interconnected factors; the factors include duration of exposure, contaminant's concentration, physicochemical properties of pollutants (e.g. solubility, pressure etc.), plants characteristics (e.g. species, root system etc.), and environmental characteristics (e.g. pH, temperature etc.).

TABLE 2.2
Recent Phytoremediation Studies Using Some Well-Known Aquatic Plants

Plant Species	Life Form	Target Contaminant	Removal Efficiencies	Ref.
Common duckweed (*Lemna minor*)	Free-floating	Methylene Blue Dye	80.56%	Imron et al., 2019a
Least duckweed (*Lemna minuta*)	Free-floating	Cr (VI) and phenol	75–85% for Cr (VI) and 100% for phenol	
Giant duckweed (*Spirodela polyrhiza*)	Free-floating	Antibiotic ofloxacin	93.73–98.36%	
		Pb	82.23–93.19%	
Water hyacinth (*Eichhornia crassipes*)	Free-floating	Ammonium nitrogen (NH_4^+-N) and dissolved organic nitrogen (DON)	>99% for both NH_4^+-N and DON	
		Cr (III)	96.70%	
Water lettuce (*Pistia stratiotes*)	Free-floating	COD, NH_4^+-N, nitrates, phosphates	47.82–88.00% for COD, 76.78–98.79% for NH_4^+-N, 16.92–97.14% for nitrates, and 73.72–92.89% for phosphates	
		Herbicide clomazone	90%	
Water milfoil (*Myriophyllum aquaticum*)	Submerged	Total phosphorus	78.2–89.8%	
Spiked water milfoil (*Myriophyllum spicatum*)	Submerged	Zinc oxide	29.5–70.3%	
		Cobalt and Cesium	90% for Co and 60% for Cs	

(Continued)

TABLE 2.2 (Continued)
Recent Phytoremediation Studies Using Some Well-Known Aquatic Plants

Plant Species	Life Form	Target Contaminant	Removal Efficiencies	Ref.
Waterthyme (*Hydrilla verticillate*)	Submerged	BOD, COD, and Suspended Solid (SS)	66.72% for BOD, 77.78% for COD, and 55.55% for SS	
		Phenol	90–99%	
Cattail (*Typha latifolia*)	Emerged	Hg, As, Pb, Cu and Zn	>80% for all metals, except for Pb 64%	
Bog bulrush (*Scirpus mucronatus*)	Emerged	Total Petroleum Hydrocarbon	74.9–82.1%	Almansoory et al., 2020
Giant bulrush (*Scirpus grossus*)	Emerged	COD, color, and SS	66.1% for COD, 55.8% for color and 87.2% for SS	
		TSS, COD, and BOD	98% for TSS, 88% for COD and 93% for BOD	
Softstem bulrush (*Scirpus validus*)	Emerged	Decabromodiphenyl ether (BDE-209, C12OBr10)	72.22–92.84%	
Common reed (*Phragmites australis*)	Emerged	Pharmaceuticals bezafibrate and paroxetine	47–75% for bezafibrate and 65–95% for paroxetine	
		Cadmium, lead, and nickel	93% for Cd, 95% for Pb, and 84% for Ni	

There have been extensive works on the application of aquatic plants in phytoremediation that focus on the ability of individual species. Meanwhile, studies that explore the ability of mixed-plant species to degrade contaminants are limited. Several studies that emphasize using plant communities have shown that a species' richness had a positive effect on the removal of both single and multiple contaminants, such as total phosphorus, BOD and metals (Pb, Cd, and Zn), and total inorganic nitrogen.

However, competition between plants should be understood as this may impact the effectiveness of contaminant removal. Moreover, a study by Geng et al. also suggests that the composition of appropriate plant species might be more important than increasing species richness. Therefore, further studies to find optimal plant combinations for the removal of particular contaminants are required since this would help optimize phytoremediation efficiency.

2.6 INDUSTRIAL WASTEWATER TREATMENT USING CONSTRUCTED WETLAND

A constructed wetland is the most used phytoremediation model that follows the basic principle of phytoremediation. A constructed wetland (CW) is divided into

TABLE 2.3
Industrial Wastewater Treatment Using Constructed Wetland

No	Industry	Wastewater	Treatment	Type of CW	Plant	OLR	HLR (cm d^{-1})	HRT (days)	% removal (%)	Year	Country	
1	Glass industry	Wastewater from washing glass sheets and the factory's machines production.	Settling tank-CW	HSFCW	pampas grass	-	-	6,8	Removal for BOD_5: 90%, COD: 90%, TSS: 99%, TN: 95%, TP: 96%	2018	Iran	Full scale Capacity 10 m^3/day
2	Tannery Industry			HSSFCW	Canna indica, T. latifolia, P. australis, Stenotaph- rum secundatum and I. pseudacorus	COD: 332–1602 kg $ha^{-1}d^{-1}$ BOD_5: 218–780 kg ha^{-1} d^{-1}	3 & 6		COD: 41–73% BOD_5: 41–58%		Portugal	Five parallel pilot units. Surface area: 1.2 m^2, depth: 0.60 m
2	Tannery Industry		Chemical-physical-CW	HSFCW (2 stages)	Phragmites australis, Typha latifolia	COD: 242 - 1925 kg $ha^{-1}d^{-1}$ and BOD: 126-900 kg ha^{-1} d^{-1}	6	2,5 & 7	BOD_5: 88% COD: 92%	2005–2006	Portugal	Onsite 2 pilot units. Surface area: 1.2 m^2, depth: 0.60 m
3	Sugar industry	Molases after Anaerobic	Anaerobic pond-CW	SFCW	Cyperus involucratus, Typha augustifolia and Thalia dealbata	BOD5: 612 kgha-1 day-1	-	-	SS: 90–93%, BOD5: 88–89%, COD: 67%, total phosphorus: 70–76%, N-NH4+: 77–82%, NO3-: 94–95% and molases pigment: 72–77%	2007	Thailand	Lab 0.6×2×0.5 m

(Continued)

TABLE 2.3 (Continued)
Industrial Wastewater Treatment Using Constructed Wetland

No	Industry	Wastewater	Treatment	Type of CW	Plant	OLR	HLR (cm d^{-1})	HRT (days)	% removal (%)	Year	Country	
4	Winery industry		Anaerobic treatment-CW	HSFCW	Typha latifolia, Phragmites australis, Elodea canadensis, Ceratophyllum demersum, Nymphaea alba and Nymphaea rustica	-		-	BOD_5: 92–98% COD: 87–98% TSS: 70–90%, total nitrogen (TN): 50–90% total phosphorus (TP): 20–60%.	2001	Italy	Onsite in 3 places: • Casa Vinicola Luigi Cecchi & Sons (Siena) • Azienda Vitivinicola "Tenuta dell'Ornellaia" (Leghorn) • Azienda Agricola La Croce (Siena)
	Winery industry	wastewater from the winery mixed with the sewage	a pre-treatment (coarse screening-Imhoff tank -equalization tank) – multistageCW	VSSFCW (140 m^2) - HSSFC-W (60 m^2) and FSFCW (30 m^2)	1. Phragmites australis L 2. Cyperus Papyrus var. Siculus, Canna indica L, 3. Scirpus lacustris L., Nymphaea alba L., Iris pseudacorus L.	-	-	110 h	removal of about 69% for TSS, 78% for COD, and 81% for BOD5	March 2014 to June 2018	Sicily (Italy)	multistage pilot CW system

Dairy industry	Reject water from dewatering aerobic sludge	Retention tank-CW	SSVFCW	Phragmites australis	BOD 13.2 g. m^{-2} d^{-1}, N-NH_4^+ 2.6 g.$m^{-2}d^{-1}$	-	-	BOD5: 88.1–90.5%, TKN: 82.4–76.5%, N-$NH4^+$: 89.2–85.7%, TP: 30.2–40.6%	2012	Poland	Pilot scale 2 bed pararel a. 10 m^2 × 0.65 m depth b. 5 m^2 × 1m depth
Olive mill industry	Effluent from extraction processs	Trickling filter - CW	VFCW	Phragmites australis	COD 88–6589 $gm^{-2}d^{-1}$; phenols 17 - 997 $gm^{-2}d^{-1}$; TKN 3.0–175 $gm^{-2}d^{-1}$; OP 3.0 - 20.0 $gm^{-2}d^{-1}$,	-	-	removals of about 70%, 70%, 75% and 87% for COD, phenols, TKN and ortho-phosphate	2010	Greece	Pilot scale of two series. each 4 units Dimensions were 96 cm × 38.5 cm × 31 cm in depth
Metalurgic industry	Effluent from production plus sewage	Primary treatment -CW	FWS CW	E. crassipes, T. domingensis and P. cor- data	-	-	7–12 days	Cr: 86%, Ni: 67%, Fe: 95%, nitrate: 70% and nitrite: 60%	2002 - 2004	Santo Tom´e (Argent-ina)	50 m length by 40 m wide and 0.5–0.8 m deep
Mining industry (gold mining)	Effluents from the mining (Hg: 0.11 ± 0.03 µg mL−1) and spiked with $HgNO_3$ (1.50 ± 0.09 µg mL−1)	Tank - CW	HSSFCW	Limnocharis flava	-	-	5 days	Hg: 90%	2016	Colombia	Lab pilot scale, four trays of 50 × 20 × 20 cm
Oil well	produced waters from oil fields (i.e., waters that have been in contact with oil in situ)	RO-CW	HSSFCW	Typha latifolia Scirpus californicus	-	-	5 day	CW decrease water soluble toxic fraction that suitable for irrigation	2000	South Carolin-a, US	Lab pilot scale 4 units of 0.19 m^2 × 0.28 m

(*Continued*)

TABLE 2.3 (Continued)
Industrial Wastewater Treatment Using Constructed Wetland

No	Industry	Wastewater	Treatment	Type of CW	Plant	OLR	HLR (cm d^{-1})	HRT (days)	% removal (%)	Year	Country	
	Industrial estate	Industrial wastewater (textile, chemicals, ghee and cooking oil, marble, steel, plastic, soap and detergent industries)	FSF-CW Free surface flow wetland	CW	T. latifolia, P. stratiotes, P. australis, C. aquatilis and A. plantago-aquatica	-	-	40 h	Pb: 50%, Cd: 91.9%, Fe: 74.1%, Ni: 40.9%, Cr: 89%, and Cu: 48.3%.	2003–2004.	Gadoon Amazai Industrial Estate (GAIE), Swabi, Pakistan	7 cells CWs with a total area of 4145.71 m^2, total capacity of 1305.58 m^3

two basic principles, free-water surface flow constructed wetland (FWSCW) and sub-surface flow constructed wetland (SSFCW). Subsurface flow is divided into vertical flow (VF) CW, horizontal flow (HF) CW, french vertical flow (FVF), CW and hybrid type CW (Parde et al., 2020). Constructed wetlands can remove high amounts of organic pollutants, especially nutrients, such as nitrogen and phosphorous. In integrated systems of wastewater treatment plants, constructed wetlands could be placed after biological secondary treatment (i.e., activated sludge system) to enhance the quality of the effluent. Table 2.3 reviews full scale constructed wetland application in industries.

2.7 CONCLUSION AND OUTLOOKS

Phytoremediation is one of the oldest techniques for removing pollutants from the environment, particularly in water and soil. The basic principle of phytoremediation is using interaction between plant roots and root microorganisms. Deep knowledge about microbe-root plant-interaction mechanisms are required to develop more robust, effective, and efficient models. Constructed wetland is the most used phytoremediation model. This model has many potential use in the future due to its robustness and flexibility. Nowadays, many advance technologies, such as microbial fuel cell (MFC), could be integrated in the constructed wetland system. The possibility of system integration between phytoremediation and another advance technology should be explored extensively to enhance the effluent quality and reduce the cost.

REFERENCES

Abdullah, S.R.S., Al-Baldawi, I.A., Almansoory, A.F., Purwanti, I.F., Al-Sbani, N.H., and Sharuddin, S.S.N. (2020). Plant-assisted remediation of hydrocarbons in water and soil: Application, mechanisms, challenges and opportunities. *Chemosphere*, *247*, 125932. 10.1016/j.chemosphere.2020.125932

Ahmad, J., Abdullah, S.R.S., Hassan, H.A., Rahman, R.A.A., and Idris, M. (2017). Screening of tropical native aquatic plants for polishing pulp and paper mill final effluent. *Malaysian J. Anal. Sci.*, *21*, 105–112. 10.17576/mjas-2017-2101-12

Al-Ajalin, F.A.H., Idris, M., Abdullah, S.R.S., Kurniawan, S.B., and Imron, M.F. (2020a). Effect of wastewater depth to the performance of short-term batching-experiments horizontal flow constructed wetland system in treating domestic wastewater. *Environ. Technol. Innov.*, *20*, 101106. 10.1016/j.eti.2020.101106

Al-Ajalin, F.A.H., Idris, M., Abdullah, S.R.S., Kurniawan, S.B., and Imron, M.F. (2020b). Evaluation of short-term pilot reed bed performance for real domestic wastewater treatment. *Environ. Technol. Innov.*, *20*, 101110. 10.1016/j.eti.2020.101110

Al-Ajalin, F.A.H., Idris, M., Abdullah, S.R.S., Kurniawan, S.B., and Imron, M.F. (2020c). Design of a reed bed system for treatment of domestic wastewater using native plants. *J. Ecol. Eng.*, *21*, 22–28. 10.12911/22998993/123256

Al-Baldawi, I.A., Abdullah, S.R.S., Anuar, N., Suja, F., and Mushrifah, I. (2015). Phytodegradation of total petroleum hydrocarbon (TPH) in diesel-contaminated water using Scirpus grossus. *Ecol. Eng.*, *74*, 463–473. 10.1016/j.ecoleng.2014.11.007

Almansoory, A.F., Idris, M., Abdullah, S.R.S., Anuar, N., and Kurniawan, S.B. (2020). Response and capability of Scirpus mucronatus (L.) in phytotreating petrol-contaminated soil. *Chemosphere*, *269*, 128760. 10.1016/j.chemosphere.2020.128760

Alshekhli, A.F., Hasan, H.A., Muhamad, M.H., and Sheikh Abdullah, S.R. (2020). Development of adsorbent from phytoremediation plant waste for methylene blue removal. *J. Ecol. Eng.*, *21*, 207–215. 10.12911/22998993/126873

Armstrong, J., Armstrong, W., and Beckett, P.M. (1992). Phragmites australis: Venturi- and humidity-induced pressure flows enhance rhizome aeration and rhizosphere oxidation. *New Phytol.*, *120*, 197–207. 10.1111/j.1469-8137.1992.tb05655.x

Backer, R., Rokem, J.S., Ilangumaran, G., Lamont, J., Praslickova, D., Ricci, E., Subramanian, S., and Smith, D.L. (2018). Plant growth-promoting rhizobacteria: Context, mechanisms of action, and roadmap to commercialization of biostimulants for sustainable agriculture. *Front. Plant Sci.*, *9*, 1473. 10.3389/fpls.2018.01473

Bansal, S., Lishawa, S.C., Newman, S., Tangen, B.A., Wilcox, D., Albert, D., Anteau, M.J., Chimney, M.J., Cressey, R.L., DeKeyser, E., Elgersma, K.J., Finkelstein, S.A., Freeland, J., Grosshans, R., Klug, P.E., Larkin, D.J., Lawrence, B.A., Linz, G., Marburger, J., Noe, G., Otto, C., Reo, N., Richards, J., Richardson, C., Rodgers, L.R., Schrank, A.J., Svedarsky, D., Travis, S., Tuchman, N., and Windham-Myers, L. (2019). Typha (Cattail) invasion in North American Wetlands: Biology, regional problems, impacts, ecosystem services, and management. *Wetlands*, *39*, 645–684. 10.1007/s13157-019-01174-7

Bolan, N.S., Park, J.H., Robinson, B., Naidu, R., and Huh, K.Y. (2011). Phytostabilization. A green approach to contaminant containment. *Adv. Agron.*, *112*, 145–204. 10.1016/B978-0-12-385538-1.00004-4

Correa, D.F., Beyer, H.L., Fargione, J.E., Hill, J.D., Possingham, H.P., Thomas-Hall, S.R., and Schenk, P.M. (2019). Towards the implementation of sustainable biofuel production systems. *Renew. Sustain. Energy Rev.*, *107*, 250–263. 10.1016/j.rser.2019.03.005

Dakora, F.D., and Phillips, D.A. (2002). Root exudates as mediators of mineral acquisition in low-nutrient environments. *Plant Soil*, *245*, 35–47. 10.1023/A:1020809400075

Das, S.K., Ghosh, G.K., and Avasthe, R. (2021). Applications of biomass derived biochar in modern science and technology. *Environ. Technol. Innov.*, *21*, 101306. 10.1016/j.eti.2020.101306

Diacono, M., Persiani, A., Testani, E., Montemurro, F., and Ciaccia, C. (2019). Recycling agricultural wastes and by-products in organic farming: Biofertilizer production, yield performance and carbon footprint analysis. *Sustainability*, *11*, 3824. 10.3390/su11143824

Elias, S.H., Mohamed, M., Nor-Anuar, A., Muda, K., Hassan, M.A.H.M., Othman, M.N., and Chelliapan, S. (2014). Water hyacinth bioremediation for ceramic industry wastewater treatment-application of rhizofiltration system. *Sains Malaysiana*.

Epa, U.S. (2019). Introduction to phytoremediation 1–7. 10.4018/978-1-5225-9016-3.ch001

Hawrot-Paw, M., Ratomski, P., Mikiciuk, M., Staniewski, J., Koniuszy, A., Ptak, P., and Golimowski, W. (2019). Pea cultivar Blauwschokker for the phytostimulation of biodiesel degradation in agricultural soil. *Environ. Sci. Pollut. Res.*, *26*, 34594–34602. 10.1007/s11356-019-06347-9

Imron, M.F., Kurniawan, S.B., Ismail, N.'Izzati, and Abdullah, S.R.S. (2020). Future challenges in diesel biodegradation by bacteria isolates: A review. *J. Clean. Prod.*, *251*, 119716. 10.1016/j.jclepro.2019.119716

Imron, M.F., Kurniawan, S.B., and Soegianto, A. (2019a). Characterization of mercury-reducing potential bacteria isolated from Keputih non-active sanitary landfill leachate, Surabaya, Indonesia under different saline conditions. *J. Environ. Manage*, *241*, 113–122. 10.1016/j.jenvman.2019.04.017

Imron, M.F., Kurniawan, S.B., Soegianto, A., and Wahyudianto, F.E. (2019b). Phytoremediation of methylene blue using duckweed (Lemna minor). *Heliyon*, *5*, e02206. 10.1016/j.heliyon.2019.e02206

Ismail, N.'Izzati, Abdullah, S.R.S., Idris, M., Hasan, H.A., Halmi, M.I.E., Al Sbani, N.H., and Jehawi, O.H. (2019). Simultaneous bioaccumulation and translocation of iron and aluminium from mining wastewater by Scirpus grossus. *Desalin. Water Treat.*, *163*, 133–142. 10.5004/dwt.2019.24201

Ismail, N.I., Abdullah, S.R.S., Idris, M., Kurniawan, S.B., Effendi Halmi, M.I., AL Sbani, N.H., Jehawi, O.H., and Hasan, H.A. (2020). Applying rhizobacteria consortium for the enhancement of Scirpus grossus growth and phytoaccumulation of Fe and Al in pilot constructed wetlands. *J. Environ. Manage.*, *267*, 110643. 10.1016/j.jenvman.2020.110643

Jehawi, O.H., Abdullah, S.R.S., Kurniawan, S.B., Ismail, N. 'Izzati I., Idris, M., Al Sbani, N.H., Muhamad, M.H., Hasan, H.A., Abu Hasan, H., Idris, M., Al Sbani, N.H., Muhamad, M.H., and Hasan, H.A. (2020). Performance of pilot hybrid reed bed constructed wetland with aeration system on nutrient removal for domestic wastewater treatment. *Environ. Technol. Innov.*, *19*, 100891. 10.1016/j.eti.2020.100891

Kadir, A.A., Abdullah, S.R.S., Othman, B.A., Hasan, H.A., Othman, A.R., Imron, M.F., Ismail, N. 'Izzati, and Kurniawan, S.B. (2020). Dual function of Lemna minor and Azolla pinnata as phytoremediator for palm oil mill effluent and as feedstock. *Chemosphere*, *259*, 127468. 10.1016/j.chemosphere.2020.127468

Kagalkar, A.N., Jadhav, M.U., Bapat, V.A., and Govindwar, S.P. (2011). Phytodegradation of the triphenylmethane dye Malachite Green mediated by cell suspension cultures of Blumea malcolmii Hook. *Bioresour. Technol.*, *102*, 10312–10318. 10.1016/j.biortech.2011.08.101

Kamaruzzaman, M.A., Abdullah, S.R.S., Hasan, H.A., Hassan, M., Idris, M., and Ismail, N. (2019). Potential of hexavalent chromium-resistant rhizosphere bacteria in promoting plant growth and hexavalent chromium reduction. *J. Environ. Biol.*, *40*, 427–433. 10.22438/jeb/40/3(si)/sp-03

Karpowicz, M., Zieliński, P., Grabowska, M., Ejsmont-Karabin, J., Kozłowska, J., and Feniova, I. (2020). Effect of eutrophication and humification on nutrient cycles and transfer efficiency of matter in freshwater food webs. *Hydrobiologia*, *847*, 2521–2540. 10.1007/s10750-020-04271-5

Kis, M., Sipka, G., and Maróti, P. (2017). Stoichiometry and kinetics of mercury uptake by photosynthetic bacteria. *Photosynth. Res.*, *132*, 197–209. 10.1007/s11120-017-0357-z

Kurniawan, S.B., Abdullah, S.R.S., Imron, M.F., Said, N.S.M., Ismail, N. 'Izzati, Hasan, H.A., Othman, A.R., and Purwanti, I.F. (2020). Challenges and opportunities of biocoagulant/bioflocculant application for drinking water and wastewater treatment and its potential for sludge recovery. *Int. J. Environ. Res. Public Health*, *17*, 1–33. 10.3390/ijerph17249312

Kurniawan, S.B., Purwanti, I.F., and Titah, H.S. (2018). The effect of pH and aluminium to bacteria isolated from aluminium recycling industry. *J. Ecol. Eng.*, *19*, 154–161. 10.12911/22998993/86147

Kwoczynski, Z., and Čmelík, J. (2021). Characterization of biomass wastes and its possibility of agriculture utilization due to biochar production by torrefaction process. *J. Clean. Prod.*, *280*, 124302. 10.1016/j.jclepro.2020.124302

Logeshwaran, P., Megharaj, M., Chadalavada, S., Bowman, M., and Naidu, R. (2018). Petroleum hydrocarbons (PH) in groundwater aquifers: An overview of environmental fate, toxicity, microbial degradation and risk-based remediation approaches. *Environ. Technol. Innov.*, *10*, 175–193. 10.1016/j.eti.2018.02.001

Mao, C., Feng, Y., Wang, X., and Ren, G. (2015). Review on research achievements of biogas from anaerobic digestion. *Renew. Sustain. Energy Rev.*, *45*, 540–555. 10.1016/j.rser.2015.02.032

McCutcheon, S.C., and Jørgensen, S.E. (2018). Phytoremediation. *Encycl. Ecol.*, *4*, 568–582. 10.1016/B978-0-444-63768-0.00069-X

Miranda, A.F., Kumar, N.R., Spangenberg, G., Subudhi, S., Lal, B., and Mouradov, A. (2020). Aquatic plants, Landoltia punctata, and Azolla filiculoides as bio-converters of wastewater to biofuel. *Plants*, *9*, 437. 10.3390/plants9040437

Mostafa, M. (2015). Waste water treatment in textile industries-the concept and current removal technologies. *Mostafa*.

Mustafa, H.M., and Hayder, G. (2020). Recent studies on applications of aquatic weed plants in phytoremediation of wastewater: A review article. *Ain Shams Eng. J.* 10.1016/j.asej.2020.05.009

Nottingham, A.T., Hicks, L.C., Ccahuana, A.J.Q., Salinas, N., Bååth, E., and Meir, P. (2018). Nutrient limitations to bacterial and fungal growth during cellulose decomposition in tropical forest soils. *Biol. Fertil. Soils*, *54*, 219–228. 10.1007/s00374-017-1247-4

Parde, D., Patwa, A., Shukla, A., Vijay, R., Killedar, D.J., and Kumar, R. (2020). A review of constructed wetland on type, treatment and technology of wastewater. *Environ. Technol. Innov.*, *21*, 101261. 10.1016/j.eti.2020.101261

Park, S., Kim, K.S., Kim, J.T., Kang, D., and Sung, K. (2011). Effects of humic acid on phytodegradation of petroleum hydrocarbons in soil simultaneously contaminated with heavy metals. *J. Environ. Sci.*, *23*, 2034–2041. 10.1016/S1001-0742(10)60670-5

Purwanti, I.F., Kurniawan, S.B., Ismail, N. 'Izzati, Imron, M.F., and Abdullah, S.R.S. (2019). Aluminium removal and recovery from wastewater and soil using isolated indigenous bacteria. *J. Environ. Manage.*, *249*, 109412. 109412.10.1016/j.jenvman.2019.109412

Purwanti, I.F., Obenu, A., Tangahu, B.V., Kurniawan, S.B., Imron, M.F., and Abdullah, S.R.S. (2020). Bioaugmentation of Vibrio alginolyticus in phytoremediation of aluminium-contaminated soil using Scirpus grossus and Thypa angustifolia. *Heliyon*, *6*, e05004. 10.1016/j.heliyon.2020.e05004

Purwanti, I.F., Simamora, D., and Kurniawan, S.B. (2018a). Toxicity test of tempe industrial wastewater on cyperus rotundus and scirpus grossus. *Int. J. Civ. Eng. Technol.*, *9*, 1162–1172.

Purwanti, I.F., Titah, H.S., Tangahu, B.V., and Kurniawan, S.B. (2018b). Design and application of wastewater treatment plant for "pempek" food industry, Surabaya, Indonesia. *Int. J. Civ. Eng. Technol.*, *9*, 1751–1765.

Purwanti, I.F.I.F., Tangahu, B.V.B.V., Titah, H.S.H.S., and Kurniawan, S.B.S.B. (2019). Phytotoxicity of aluminium contaminated soil to scirpus grossus and typha angustifolia. *Ecol. Environ. Conserv.*, *25*, 523–526.

Rahman, M.A., and Hasegawa, H. (2011). Aquatic arsenic: Phytoremediation using floating macrophytes. *Chemosphere*, *83*, 633–646. 10.1016/j.chemosphere.2011.02.045

Rezania, S., Oryani, B., Cho, J., Sabbagh, F., Rupani, P.F., Talaiekhozani, A., Rahimi, N., and Ghahroud, M.L. (2020). Technical aspects of biofuel production from different sources in Malaysia—A review. *Processes*. 10.3390/PR8080993

Sandhi, A., Landberg, T., and Greger, M. (2018). Effect of pH, temperature, and oxygenation on arsenic phytofiltration by aquatic moss (Warnstorfia fluitans). *J. Environ. Chem. Eng.*, *6*, 3918–3925. 10.1016/j.jece.2018.05.044

Schwammberger, P.F., Lucke, T., Walker, C., and Trueman, S.J. (2019). Nutrient uptake by constructed floating wetland plants during the construction phase of an urban residential development. *Sci. Total Environ.*, *677*, 390–403. 10.1016/j.scitotenv.2019.04.341

Shahid, M.J., AL-surhanee, A.A., Kouadri, F., Ali, S., Nawaz, N., Afzal, M., Rizwan, M., Ali, B., and Soliman, M.H. (2020). Role of microorganisms in the remediation of wastewater in floating treatment wetlands: A review. *Sustainability*, *12*, 5559. 10.3390/su12145559

Sharma, P., Tripathi, S., Chaturvedi, P., Chaurasia, D., and Chandra, R. (2021). Newly isolated Bacillus sp. PS-6 assisted phytoremediation of heavy metals using Phragmites communis: Potential application in wastewater treatment. *Bioresour. Technol.*, *320*, 124353. 10.1016/j.biortech.2020.124353

Sharuddin, S.S.N.B., Abdullah, S.R.S., Hasan, H.A., and Othman, A.R. (2018). Comparative tolerance and survival of scirpus grossus and lepironia articulata in real crude oil sludge. *Prog. Color. Color. Coatings*. 10.14419/ijet.v8i1.16522

Sun, G., Zhao, Y.Q., and Allen, S.J. (2007). An alternative arrangement of gravel media in tidal flow reed beds treating pig farm wastewater. *Water. Air. Soil Pollut.* 10.1007/s11270-006-9316-6

Tangahu, B.V., Ningsih, D.A., Kurniawan, S.B., and Imron, M.F. (2019). Study of BOD and COD removal in batik wastewater using Scirpus grossus and Iris pseudacorus with intermittent exposure system. *J. Ecol. Eng.*, *20*, 130–134. 10.12911/22998993/105357

Titah, H.S., Purwanti, I.F., Tangahu, B.V., Kurniawan, S.B., Imron, M.F., Abdullah, S.R.S., and Ismail, N. 'Izzati (2019). Kinetics of aluminium removal by locally isolated Brochothrix thermosphacta and Vibrio alginolyticus. *J. Environ. Manage.*, *238*, 194–200. 10.1016/j.jenvman.2019.03.011

Titah, H.S., Rozaimah, S., Abdullah, S.R.S., Idris, M., Anuar, N., Basri, H., Mukhlisin, M., Tangahu, B.V., Purwanti, I.F., and Kurniawan, S.B. (2018). Arsenic resistance and biosorption by isolated Rhizobacteria from the roots of Ludwigia octovalvis. *Int. J. Microbiol.*, *2018*, 1–10. 10.1155/2018/3101498

Ullah, A., Heng, S., Munis, M.F.H., Fahad, S., and Yang, X. (2015). Phytoremediation of heavy metals assisted by plant growth promoting (PGP) bacteria: A review. *Environ. Exp. Bot.*, *117*, 28–40. 10.1016/j.envexpbot.2015.05.001

Varma, M., Gupta, A.K., Ghosal, P.S., and Majumder, A. (2021). A review on performance of constructed wetlands in tropical and cold climate: Insights of mechanism, role of influencing factors, and system modification in low temperature. *Sci. Total Environ.*, *755*, 142540. 10.1016/j.scitotenv.2020.142540

Wang, X., Fernandes de Souza, M., Li, H., Tack, F.M.G., Ok, Y.S., and Meers, E. (2020). Zn phytoextraction and recycling of alfalfa biomass as potential Zn-biofortified feed crop. *Sci. Total Environ.*, *760*, 143424. 10.1016/j.scitotenv.2020.143424

3 Microalgae as Biological Cleanser for Wastewater Treatment

Sachin Palekar, Sandhya Menon, and Nandini Girish
Department of Bioanalytical Sciences, Ramnarain Ruia Autonomous College, Mumbai, Maharashtra, India

Nilesh S. Wagh and Jaya Lakkakula
Amity Institute of Biotechnology, Amity University Maharashtra, Mumbai, Maharashtra, India

CONTENTS

3.1 INTRODUCTION

Algae are photosynthetic autotrophs that grow in various natural and non-natural aquatic environments like ponds, lakes, rivers, oceans, and effluent water. They are believed to contribute over 50% of the earth's oxygen demands. These

DOI: 10.1201/9781003165101-3

organisms are capable of growing in strained environments and hence can sustain diverse physicochemical changes such as temperature, pH, salinity, availability, and quality of light, as well as scarcity of nutrients. The key habitats are aquatic; however, it is not unusual to find algae on terrestrial surfaces or in wastewater, such as municipal wastewater or industrial wastewater. They mostly grow alone or sometimes in symbiotic associations with other organisms, e.g., lichens. (Barsanti, 2008).

Based on size, algae can be classified as macroalgae and microalgae. Macroalgae are multicellular, non-microscopic alga most commonly known as seaweeds. On the other hand, microalgae are unicellular, microscopic organisms (Muhammad Imran Khan, 2018).

Microalgae is a chief group of phytoplanktons having enormous diversity of prokaryotic cyanobacteria and eukaryotic photoautotrophs. Microalgae include major types like green algae, brown algae, red algae, and blue-green algae, to name a few. The microalgal diversity is displayed in morphology and habitats. Prokaryotic microalgae, like chloroxybacteria, are similar to cyanobacteria, whereas eukaryotes are closer to chlorophyceae members (Corrêa, 2017).

Algae are known to be useful in many ways and are mass produced for various purposes. Some uses of algae include functional foods and food additives, fodder for livestock, fertilizer, a source of dyes and pigments, biofertilizers, biofuel, polysaccharides, pollution control, and bioremediation, to name a few. Microalgae are potential reserves of commercially important carbon complexes that can be used for developing alternatives to nonrenewable natural resources (Das, 2011).

The fact that algae, specifically microalgae, survive in many unusual and difficult-to-survive habitats, like polluted water bodies, makes them a potential candidate for treating such contaminated water sources, rendering them suitable for agriculture and other allied activities (Muhammad Imran Khan, 2018).

Before addressing the decontamination of the polluted water sources, it is imperative to discuss the sources of contaminants and pollutants contributing to wastewater. Around 70% of the earth's surface is covered with water. This water is distributed between the ocean bodies, freshwater, and underground water. Approximately 3% of the total water is either trapped as glaciers and icecaps or available as freshwater for immediate use through rivers and lakes or stored as underground freshwater. This data suggests that of all the water available on earth, less than 0.5% is available for use (Christenson, 2011; Abdel-Raouf, 2012; Rachel Whitton, 2015).

Unfortunately, with industrialization, many water bodies, freshwater as well as saline, are getting contaminated. This contamination would ultimately affect all species on earth and might even pose a risk of the extinction of certain species (Hoffmann, 1998).

The agents that pollute these water bodies can be either biological or chemical. For instance, pathogens released from untreated sewage water are biological pollutants. Industrial waste, fertilizers, and pesticides used for agriculture, oil spills in oceans, or fuels like petroleum hydrocarbons form the examples of chemical pollutants. These pollutants not only affect the availability of these water sources for use but also the flora and fauna surviving near or in these water bodies (Iyyanki V. Muralikrishna, 2017). Treating this wastewater therefore becomes imperative for the survival of all species. Wastewater can be treated in various ways, like biological treatment, chemical treatment, and physical treatment. Chemical treatment is practiced by treating sewage or industrial runoffs using chemicals like chlorine, ozone, or neutralizers like lime. Physical treatment methods involve using techniques like sedimentation or filtration for removing impurities in the water. Biological treatment involves using some biological agent to decontaminate the water. Microalgae are promising candidates for this purpose (Hoffmann, 1998).

Microalgae have several advantages over chemical and physical agents for water decontamination. They not only can decontaminate sewage water but can be used for industrial wastewater treatment. Their use for treating wastewater and hazardous contaminants is specifically advantageous due to their ability to uptake minerals, ions, and organic compounds (Zeng, 2012). They can also be used for removing heavy metals in wastewater. Being autotrophic and known to use certain unusual metabolites, such as hydrocarbons as a source of nutrition, they can be counted as a cheap and perfect candidate for wastewater treatment (Sydney, 2011; El-Ghonemy, 2012).

This chapter emphasizes the use of microalgae as a viable and sustainable agent for phycoremediation. Additionally, this chapter also provides a detailed account of microalgae accomplished for eliminating the harmful substances such as hydrocarbons, heavy metals, and pesticides using diverse methods like bioconcentration, biotransformation, volatilization, and biosorption. The chapter also discusses challenges associated with optimizing the yield, contamination and reducing the resilience time of the algal culture. Additionally, the chapter provides an insight into the methods for genetic modification of microalgae for their targeted development as an efficient device for phycoremediation.

3.2 MICROALGAE AS PHYCOREMEDIATION AGENT

Phycoremediation is a type of bioremediation where the biotic component involved belongs to the algal kingdom. The process of phycoremediation involves the use of both microalgae and macroalgae for decontaminating the damaging chemicals from wastewater. It also includes the biotransformation or reduction of hazardous pollutants from the wastewater. It can also be applied for elimination of nutrients from wastewater containing animal waste or having higher organic content (Walter Mulbry, 2008; EJ., 2003). The employment of micro algae for revitalizing the environment is more favourable over the others because of its short life cycle and rapid metabolism and growth (Hoffmann, 1998).

Wastewaters can have their origin from several sources that can be the large-scale setups, such as the chemical industry or any other manufacturing plant, or they may originate from small-scale setups like the poultry, dairy, or fish farms. But, in all cases, these waters are variably loaded with components that need to be filtered before they are let out in the natural, larger sources of water (Molazadeh Marziyeh, 2019).

As displayed in Table 3.1, many species of micro algae can be successfully employed for treating wastewaters with a view of liberating them from their chemical, pesticidal, or heavy metal load.

As per the current developments *Scenedesmus* and *Chlorella* can be considered the most promising microalga for phycoremediation of wastewaters (Kayil Veedu Ajayan,

2015). *Arbib et al.* have promoted the use of two *Chlorella spp; Chlorella vulgaris and Chlorella kessleri,* along with *Scenedesmus obliquus* for the tertiary treatment of urban sewage. The treatment of dairy wastewaters using Chlorella spp. has also been documented by many (Zouhayr Arbib, 2012). *Scenedesmus* spp are known to have a role in the phycoremediation of toxicants from tannery wastewaters (Kayil Veedu Ajayan, 2015). Tannery wastewater is a serious concern not only because of its high organic content but also because it contains chromium originating from the chromium salts used during the tanning process. This water therefore is a very strong environmental pollutant. The chromium accumulating capacity of *Scenedesmus* spp, if employed wisely, can make a perfect equation for environment protection. In another study, barium bioaccumulating capacity of *Scenedesmus subspicatus, Selenastrum capricornutum*, and *Nannochloropsis spp* has been brought to the fore. Thus, it can be understood that microalga can play a mutualistic role with the environment by their natural property of heavy-metal accumulation (Kayil Veedu Ajayan, 2015). Another modern-day concern, especially among the urban crowd, is that of pesticide accumulation in various natural resources, which eventually finds its way into the food chain and may pose serious health hazards. Both *Chlorella* and *Scenedesus spp* have been proven to be influential in the removal of pesticides from wastewaters, which is a very beneficial feature not only for the human beings but also for the entire planet (Kayil Veedu Ajayan, 2015; Shehata, 1980)

3.3 WASTEWATER TREATMENT WITH MICROALGAE

The United Nations in the analytical brief on wastewater management defines wastewater as a combination of one or more of these: domestic effluent consisting

TABLE 3.1
Highlights Other Such Micro Alga and their Capabilities

Sr No.	Name of the Microalgae	Advantage	Reference
1	*Parachlorella kessleri-I*	Can remove pollutants from municipal and aquaculture wastewater	(Amit Kumar Singh, 2017; Liu, 2019)
2	*Arthrospira platensis (Spirulina)*	Can serve as a bioremediation agent for nickel, zinc, cadmium, and copper	(Marco Piccini, 2019)
3	*Dunaliella salina*	Can bring about degradation of radioactive atoms from aqueous solutions	(Azmoonfar, 2019)
4	*Selenastrum capricornutum*	Pesticide residue management, bioaccumulation of barium from water	(Nie, 2020)
5	*Euglena gracilis*	Pesticide residue management	(Nie, 2020)
6	*Nannochloropsis oculate*	Capable of absorbing lead, potential to remove nitrogen and phosphorus compound in wastewater, can biodegrade formaldehyde	(Dwivedi, 2012)

of blackwater (excreta, urine, and faecal sludge) and greywater (kitchen and bathing wastewater); water from commercial establishments and institutions, including hospitals; industrial effluent, stormwater, and other urban run-off; agricultural, horticultural and aquaculture effluent, either dissolved or as suspended matter. Wastewater is thus a complex mixture of natural organic and inorganic materials, as well as man-made compounds. Organic carbon in sewage is mostly present in the form of carbohydrates, fats, proteins, amino acids, and volatile acids. The inorganic constituents consist of sodium, calcium, potassium, magnesium, chlorine, sulphur, phosphate, bicarbonate, ammonium salts, and heavy metals (Abdel-Raouf, 2012). These 'waste' components are a source of nutrition for many algal species like *Scenedesmus* spp or *Parachlorella kessleri-I* to name a few. This indicates that the treatment of wastewater can be lined to production of algal biomass, biofuels, and bioproducts (Christenson, 2011).

Wastewater treatment processes can be classified into primary, secondary, and tertiary treatment processes. The type of treatment required for treating wastewater depends on the composition of wastewater. Microalgae as a method of wastewater treatment can be considered as a part of the secondary/tertiary treatment process where biological materials like microalgae are used to remove contaminants (Abdel-Raouf, 2012).

Wastewater treatment by microalgae involves fixing atmospheric carbon dioxide by the photosynthetic algae and utilizing the inorganic "nutrients" present in the wastewater (Acién, 2016). This means that the contaminants in wastewater the microalgae use ensure the removal of the toxic components. Wastewater also contains organic matter like effluent water from the food and dairy industry, effluent from the liquor industry, household waste, etc. Many algae have the ability to carry out chemoheterotrophic or mixotrophic metabolism, which can help eliminate these waste products. Utilization of wastewater from dairy, piggery, or beet vinnase by *Arthrospira platensis* or removal of nitrogen and phosphorus from municipal wastewater effluent by *Chlorella vulgaris* are a few examples of using algae to degrade wastewater (Hena, 2018; Depraetere, 2013; Coca, 2015).

Bioremediation is the ability of living organisms, mostly microorganisms, to degrade the pollutants such as heavy metals, pesticides, or other contaminants and convert them into less toxic or nontoxic forms (Jerome, 2019). For the purpose of understanding the process of phycoremediation (bioremediation using algae), let us consider the example of phosphorus and nitrogen removal by bioremediation. Both nitrogen and phosphorus are essential to the growth of all organisms, including algae. However, with excessive use of chemical fertilizers, these elements are often found in excessive amounts in water bodies, posing a serious health hazard to humans and other living organisms as well.

3.3.1 Microalgae in Removal of Nutrients from Wastewater

Microalgae serve as excellent photobioreactors for removal of nutrients from the water samples due to the light-driven metabolic processes allied with nutrient assimilation. Direct nutrient remediation is achieved through consistent biochemical trails for the utilization of the desired nutrients into the biomass for storage, or integration into nucleic acids and proteins for biomass growth ((Rachel Whitton, 2015).

Pittman *et al.* proved that algae can thrive in nitrogen- and phosphorus-rich conditions that are found in many wastewaters. This could provide a high value by-product for algae that are primarily being grown for biofuel (Pittman, 2011). When phosphorus microalgae are depleted of phosphorus in the external environment, they tend to accumulate phosphorus whenever it becomes available to them. Though not much is known about how microalgae regulate this uptake, it becomes an interesting facet of the microalgae. Aslan and Kapdan discussed that algae can remove nitrogen more effectively than phosphorus from their environment, including polluted water bodies. Microalgae absorb this nitrogen from polluted water bodies and store it as biomass, which can later be released back into the environment (Aslan, 2006). This ability of the algae to use the wastewater as a source of nutrition not only fixes the problem of treating wastewater but is also economical because many such microalgae produce products of economic importance like triacylglycerides (TAGs) or long-chain hydrocarbons (LCHs).(Niehaus, 2011).

3.3.2 Microalgae for Removal of Heavy Metals from Wastewater

Heavy metals are another example of commonly found contaminants in wastewater with significant effect on the human body; the effects vary from growth and developmental abnormalities, carcinogenesis, neuromuscular control defects, mental retardation, or renal malfunction, to state a few. Microalgae like *Scenedesmus* sp. have the capability to take up these heavy metals, such as copper, cadmium, nickel, zinc, and lead (Shehata, 1980). It is believed that these heavy metals adsorb on the cell surface of the microalgae after which they are slowly transported inside the cell. Many heavy metals such as Co, Pb, Mg, Zn, Cd, Sr, Hg, Ni and Cu are sequestered in polyphosphate bodies in green algae. These bodies are not only storage centers of such heavy metals but are also responsible for their detoxification. Studies have proved that the alga *Scenedesmus obliquus* accumulated Cd and Zn on increasing the amount of phosphorus in the media. *Cladophora glomerata* has shown removal of Cu, Pb, Cd, Co from water bodies (Dwivedi, 2012).

3.3.3 Microalgae for Removal of Pesticides from Wastewater

The use of pesticides is essential to increase crop productivity. But, the recurrent usage of these chemical agents can lead to stern ecological complications due to bioaccumulation and biomagnification. Lately, microalgae technology has become the sustainable source for the effective removal of pesticides from wastewater (Nie, 2020). The application of microalgal consortia for removal of pesticides can be seen in Figure 3.1.

Figure 3.1 is a synoptic representation of a recent study where microalgal consortia were used for elimination of hydrophobic pesticides, viz., chlorpyrifos, oxadiazon, and cypermethrin from the water-based media. Some 14% of cypermethrin, 35% of chlorpyrifos, and 55% of oxadiazon were decontaminated from water-based media by photodegradation and biodegradation. Additionally, biosorption was successfully

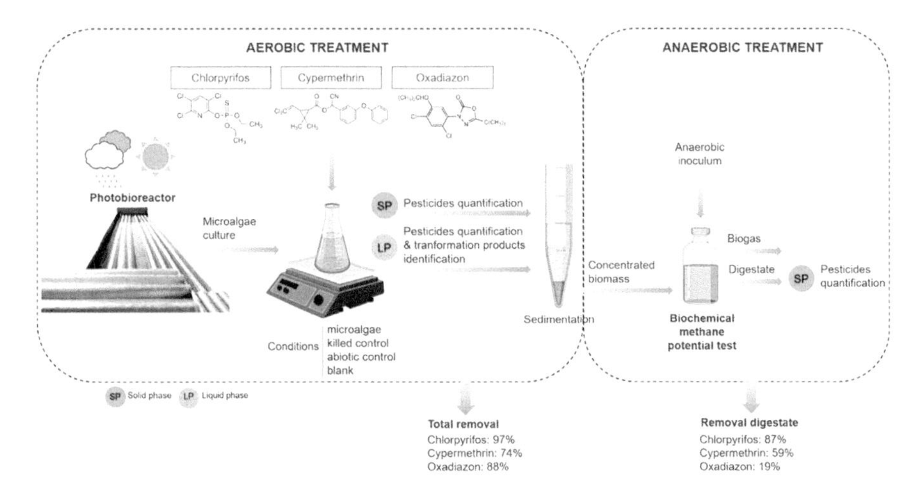

FIGURE 3.1 Microalgae in the remediation of hydrophobic pesticides. Reprinted with permission from Avila, R. P. (2021). Biodegradation of hydrophobic pesticides by microalgae: Transformation products and impact on algae biochemical methane potential. Science of the Total Environment, 754, 142114.

employed to decontaminate 60% of chlorpyrifos and cypermethrin onto microalgae biomass (Avila, 2021).

3.4 MICROALGAE CULTIVATION SYSTEMS FOR WASTEWATER TREATMENTS

Microalgae are cultivated for wastewater treatment primarily in an open system, a closed system, or a hybrid system. They are chosen on the basis of the components of wastewater, the nutrients that need to be supplemented along with the wastewater for microalgal growth, the light/pH/temperature requirements, and the cost to be incurred in setting up, running, and maintaining the system or the final products obtained from the process (Jia, 2016; Molazadeh Marziyeh, 2019).

3.4.1 Open Systems

These systems are generally classified into natural or artificial water systems. Natural water systems include lakes, lagoons, and ponds, while artificial water systems include man-made ponds, tanks, and various above-ground and below-ground containers (Ugwu, 2008). Open systems can also be classified as non-stirred and stirred ponds.

3.4.1.1 Non-Stirred Ponds

Water systems that are not or cannot be subjected to are called non-stirred ponds. Natural water systems belong to this category. Inability to stir a water system may lead to poor mixing of the algal culture, but it has the advantage of lower cost for commercial scale up. Natural lakes, lagoons, and ponds are some examples of non-stirred ponds for microalgae cultivation (Shen, 2009).

3.4.1.2 Stirred Ponds

Stirred ponds come with a system to ensure the stirring of the medium in which algal growth occurs as it ensures nutrient availability to all the cells. High-rate algal ponds (HRAP) and circular ponds are two common examples of stirred ponds (Fazal, 2018; Anto, 2020).

HRAP is also known as raceway ponds consist of paddlewheel, baffle, and channels. The raceway ponds are usually open and shallow with depth ranging from 15 cm to 50 cm. They can be constructed with either individual or multiple channels in a closed loop. The paddlewheels mix up the algae with the medium, ensuring no settling of algae, while the baffles direct the medium and cells to move through the channels. This ensures algae cells are sufficiently mixed and keep a continuous flow to receive sunlight and CO_2 from the atmosphere (Ting, 2017).

3.4.2 Closed System

Closed systems provide a controlled environment for growth of microalgae and enable a better yield of microalgae biomass. They help in overcoming the problems of contamination and environmental influence encountered within open systems but are expensive and challenging to maintain (Pacheco, 2020). They also incur a high operation and maintenance cost. A variety of reactor designs are available in this category, such as the tubular photobioreactor, flat plate photobioreactor, helical photobioreactor, and plastic bag photobioreactor. But a closed system is much more expensive and incurs a high operation and maintenance cost. The reactor designs can be varied like tubular photobioreactor, flat plate photobioreactor, helical photobioreactor, plastic bag photobioreactor (Molazadeh Marziyeh, 2019).

3.4.2.1 Flat Plate Photobioreactor

This design consists of a rectangular box appearance that can be exposed to artificial light sources indoors or to sunlight. The plates are made of transparent or semi-transparent materials like glass. They allow light to easily penetrate the culture medium. They are flexible and allow for scale up, but pose the risk of biofilm formation (Molazadeh Marziyeh, 2019; Ting, 2017).

3.4.2.2 Tubular Photobioreactors

This type of bioreactors can be vertical, horizontal, or helical tubular. The algae and growth media are continuously circulated through the tubes with the help of an airlift or mechanical pump. However, these come with a heavy operating cost and are also difficult to maintain (Molazadeh Marziyeh, 2019).

3.4.2.3 Plastic Bag Photoreactor

This design of bioreactors consists of plastic bags fitted with aerators. They are hung vertically or placed in a metal or plastic cage for support and exposed to direct sunlight. They are also difficult to maintain and require constant attention

(Ting, 2017). Whichever may be the method chosen for cultivation, it is noteworthy that microalgae are not always present as pure cultures and interact with other organisms in their community. Therefore, when using microalgae for wastewater treatment, they are mostly used in combinations with other microalgae. This co-culture method can involve either the coexistence of different types of microalgae or a concurrence of microalgae with bacteria or fungi.

Another method for wastewater treatment involves immobilizing the microalgae in polymer gel matrices. The immobilized microalgae methods allow for easier separation of biomass separation but can be expensive (Molazadeh Marziyeh, 2019).

3.5 CHALLENGES ASSOCIATED WITH CULTIVATION OF MICROALGAE

The natural, open systems may seem the most alluring choice for microalgal cultivation due to their *prima facie* inexpensive nature in comparison with the closed systems. However, one has to understand that the controlled environment of the closed systems can be a more validated setup for generating high yields of micro algae reproducibly. But the capital and recurring expenditures associated with the closed system can be a hurdle in the glorification of microalgae as cost-effective bioremediators (Ugwu, 2008; Pacheco, 2020; Molazadeh Marziyeh, 2019).

Another problem in microalgae cultivation is their tendency to co-exist with other cultures, which may have a greater probability of occurrence in the uncontrolled open environments (Pacheco, 2020; Molazadeh Marziyeh, 2019). Certain genetic modifications, which may enhance the quorum-sensing abilities or may instil or enhance biofilm-like growth capacities, may help in achieving pure microalgal cultures, even in open systems. This may also increase the global interest toward the propagation of bioremediation methods using the microalgae (Nie, 2020).

Research on the above lines may widen the scope, applications, and acceptance of microalgae as natural remediators, which may perhaps alleviate the present environmental circumstances.

3.6 GENETIC MODIFICATIONS OF MICROALGAE FOR EFFICIENT DECONTAMINATION

Considering the potential uses of microalgae, especially in treating wastewater, engineering them for efficient wastewater removal is highly recommended. Since these microalgae also have the ability to produce commercially viable products like TAGs, a genetic-engineering approach would lead not only to efficient wastewater treatment but also to a significant increase in biomass. Currently, many techniques are available that help in better understanding the genome of microalgae. This would in turn help in identifying the potential genes for manipulating for wastewater treatment. Genetic modifications in small and large subunits of Rubisco (rbcS and rbcL) to make hybrid Rubiscos that are combination of plant (rbcS) and algal subunits (rbcL) have been carried out in *C. reinhardtii*. CO_2 fixation (photosynthesis) has been studied. This helps in net increase in CO_2 fixation while also increasing the growth of these organisms (Genkov, 2010; Whitney, 2011; Guihéneuf Freddy, 2016). Such modifications can also be tried for utilization of nutrients from wastewater.

Some researchers are working toward developing better improved strains with higher ability to detoxify waterbodies of heavy metals. Scientists have reported genetic modificatons in *C.reinhardtii* to insert a foreign metallothionein (MT-II) that caused a twofold increase in the removal of cadmium as compared to wild type strains (Rehnstam-Holm, 2003). They have also suggested genetic modification in certain species of cyanobacteria to efficiently remove pesticides. Many cyanobacteria are able to degrade lindane by dechlorinating lindane with the help of *nir* operon, the operon responsible for nitrate utilization. An increase in this ability of *Anabaena* sp. and *Nostoc ellipsosorum* to degrade lindane by inserting a lindane dechlorination operon from the bacterium *Pseudomonas paucimobilis,* thereby uncoupling lindane degradation from *nir* operon, has been reported. The results suggested an enhanced phytochelatin synthesis and sequestration of Cd in Cd-stressed cells (Siripornadulsil, 2002). Rajamani (2007) also suggested engineering microalgae for heavy metal bioremediation, requiring genetic engineering and biological considerations to ensure no transgenic organisms escape; the release of live transgenic microalgae with enhanced heavy-metal binding capacities into the environment has its own disadvantages. Such transgenic algae with enhanced heavy-metal capacities could lead to these heavy metals accumulating in food chains. The genetic-engineering techniques therefore ensure multiple checkpoints to prevent the survival of such transgenic organisms in the natural environment (Rajamani, 2007).

A treatment system has been reported using *Chlamydomonas* cells engineered for enhanced metal binding capacity in closed chambers with the heavy metal containing wastewater being passed across semipermeable membranes. This ensured the entrapment of the engineered cells while decontaminating the wastewater of heavy metals. An additional mutation of pf-14 mutations or nit1–30 in such organisms would ensure that any cells escaping from the containment system are non-viable; pf-14 mutations would prevent mating due to paralyzed while nitl-30 mutations would inhibit growth on nitrate. A post-treatment sterilization of the wastewater before release would kill any living cells in the system (Rajamani, 2007; Guihéneuf Freddy, 2016).

Such transgenic strains of microalgae would be even more beneficial in recovering and purifying contaminated wastewater bodies (Rajamani, 2007). Further research is needed in identifying means to improve the strains known in phycoremediation while ensuring no harm to the ecosystem. This will help in placing algae as a primary method for treatment for cleaning such polluted water bodies.

3.7 CONCLUSION

Microalgae has proven to be one of the vital tools for removal of heavy metals, nutrients, and micropollutants from wastewater originating from diverse sources as sewage, industrial, agricultural, and poultry waste. Microalgae can be grown in an open system, a closed system, or a hybrid system, which can be selected based on quality and components of wastewater. Mostly microalgal consortia is used where the synergy of these microalgal organisms efficiently cleanse or decontaminate the water. Therefore, diverse and useful substances are obtained as by-products. However, standardization of existing techniques and development and validation of newer methods with greater yields for microalgal cultivation will help in the adoration of this humble, eco-friendly strategy for the cleansing of wastewater.

REFERENCES

Abdel-Raouf, A. (2012). Microalgae and wastewater treatment. *Saudi Journal of Biological Sciences*, *19*(3), 257–275.

Acién, F.G., Gómez-Serrano, C., Morales-Amaral, M.M., Fernández-Sevilla, J.M., Molina-Grima, E. (2016). Wastewater treatment using microalgae: How realistic a contribution might it be to significant urban wastewater treatment? *Applied Microbiology and Biotechnology*, *100*(21), 9013–9022.

Amit Kumar Singh, N.S. (2017). Phycoremediation of municipal wastewater by microalgae to produce biofuel. *International Journal of Phytoremediation*, *19*(9), 805–812. doi: 10.1080/15226

Anto, S.M. (2020). Algae as green energy reserve: Technological outlook on biofuel production. *Chemosphere*, *242*, 125079.

Aslan, S.K. (2006). Batch kinetics of nitrogen and phosphorus removal from synthetic wastewater by algae. *Ecological Engineering*, *28*, 64–70.

Avila, R.P. (2021). Biodegradation of hydrophobic pesticides by microalgae: Transformation products and impact on algae biochemical methane potential. *Science of the Total Environment*, *754*, 142114.

Azmoonfar, R.N. (2019). Adsorption of radioactive materials by green Microalgae Dunaliella Salina From Aqueous solution. *Iranian Journal of Medical Physics*, *16*(9), 392–396.

Barsanti, L. (2008). Oddities and curiosities in the algal world. In L.B.V. Evangelista (Ed.), *Algal toxins: Nature, occurrence, effect and detection* (pp. 353–391). Springer, Dordrecht.

Christenson, L. (2011). Production and harvesting of microalgae for wastewater treatment, biofuels, and bioproducts. *Biotechnology Advances*, *29*(6), 686–702.

Coca, M. B.-B. (2015). Protein production in Spirulina platensis biomass using beet vinasse-supplemented culture media. *Food and Bioproducts Processing*, *94*, 306–312.

Corrêa, I. (2017). Deep learning for microalgae classification. *ICMLA*, 0–183.

Das, P. (2011). Two phase microalgae growth in the open system for enhanced lipid productivity. *Renew Energy*, *36*(9), 2524–2528.

Depraetere, O.F. (2013). Decolorisation of piggery wastewater to stimulate the production of Arthrospira platensis. *Bioresource Technology*, *148*, 366–372.

Duraiarasan, S. (2012). A comprehensive review on the potential and alternative biofuel. *Research Journal of Chemical Sciences*, *2*, 71–82.

Dwivedi, S. (2012). Bioremediation of heavy metal by algae: Current and future perspective. *Journal of Advanced Laboratory Research in Biology*, *3*(3), 195–199.

El-Ghonemy, A. (2012). Future sustainable water desalination technologies for the Saudi Arabia: A review. *Renewable and Sustainable Energy Reviews*, *16*, 6566–6597.

Fazal, T.M. (2018). Bioremediation of textile wastewater and successive biodiesel production using microalgae. *Renewable and Sustainable Energy Reviews*, *82*, 3107–3126.

Genkov, T.M. (2010). Functional hybrid rubisco enzymes with plant small subunits and algal large subunits: Engineered rbcS cDNA for expression in chlamydomonas. *Journal of Biological Chemistry*, *285*, 19833–19841.

Guihéneuf Freddy, K. A.-S. (2016). Genetic engineering: A promising tool to engender physiological, biochemical, and molecular stress resilience in green microalgae. *Frontiers in Plant Science*, *7*, 400.

Hena, S.Z. (2018). Dairy farm wastewater treatment and lipid accumulation by Arthrospira platensis. *Water Research*, *128*, 267–277.

Hoffmann, J.P. (1998). Wastewater treatment with suspended and nonsuspended algae. *Journal of Phycology, 34*, 757–763.

Iyyanki V. Muralikrishna, V.M. (2017). *Chapter One - Introduction environmental management* (1–4 ed.). Iyyanki V. Muralikrishna, V.M. (Ed.). Butterworth-Heinemann. doi: 10.1016/B978-0-12-811989-1.00001-4

Jerome, N. (2019). *Encyclopedia of environmental health* (2nd ed., Vol. 3). J. Nriagu (Ed.). Elsevier, Michigan, USA.

Jia, H. (2016). Removal of nitrogen from wastewater using microalgae and microalgae–bacteria consortia. *Cogent Environmental Science, 2*(1), 1275089.

Kayil Veedu Ajayan, M.S. (2015). Phycoremediation of tannery wastewater using microalgae scenedesmus species. *International Journal of Phytoremediation, 17*(10), 907–916. doi: 10.1080/15226514.2014.989313

Liu, Y.L. (2019). Treatment of real aquaculture wastewater from a fishery utilizing phytoremediation with microalgae. *Journal of Chemical Technology & Biotechnology, 94*, 900–910.

Marco Piccini, S.R. (2019). A synergistic use of microalgae and macroalgae for heavy metal bioremediation and bioenergy production through hydrothermal liquefaction†. *Sustainable Energy Fuels, 3*, 292–301.

Molazadeh Marziyeh, A.H. (2019). The use of microalgae for coupling wastewater treatment with CO_2 biofixation. *Frontiers in Bioengineering and Biotechnology, 7*, 42.

Muhammad Imran Khan, J.H. (2018). The promising future of microalgae: Current status, challenges, and optimization of a sustainable and renewable industry for biofuels, feed, and other products. *Microbial Cell Factories, 17*(36), 1–21.

Nie, S.Y. (2020). Bioremediation of water containing pesticides by microalgae: Mechanisms, methods, and prospects for future research. *Science of the Total Environment, 707*, 136080.

Niehaus, T.O. (2011). Identification of unique mechanisms for triterpene biosynthesis in Botryococcus braunii. *PNAS, 108*, 12260–12265.

Olguín, E.J. (2003). Phycoremediation: Key issues for cost-effective nutrient removal processes. *Biotechnology Advances, 22*(1–2), 81–91.

Pacheco, D.R. (2020). Microalgae water bioremediation: Trends and hot topics. *Applied Sciences, 10*(5), 18–86.

Phang, S.M. (1988). Algal biomass production in digested palm oil mill effluent. *Biological Wastes, 25*, 77–191.

Pittman, J.D. (2011). The potential of sustainable algal biofuel production using wastewater resources. *Bioresource Technology, 102*, 17–25.

Rachel Whitton, F.O. (2015). Microalgae for municipal wastewater nutrient remediation: Mechanisms, reactors and outlook for tertiary treatment. *Environmental Technology Reviews, 4*(1), 133–148.

Rajamani, S.S. (2007). Phycoremediation of heavy metals using transgenic microalgae. *Transgenic Microalgae as Green Cell Factories*, 99–109.

Rehnstam-Holm, A.S. (2003). *Genetic engineering of algal species* (pp. 1–27). Eolss Publishers, Oxford, UK.

Scott, S.D.-S. (2010). Biodiesel from algae: Challenges and prospects. *Current Opinion in Biotechnology, 21*, 277–286.

Shehata, S.B. (1980). Growth response of Scenedesmus to differentconcentrations of copper, cadmium, nickel, zinc and lead. *Environment International, 4*, 431–434.

Shen, Y. (2009). Microalgae mass production methods. *Transactions of the ASABE, 52*, 1275–1287.

Siripornadulsil, S.T. (2002). Molecular mechanisms of proline-mediated tolerance to toxic heavy metals in transgenic microalgae. *Plant Cell, 14*, 2837–2847.

Sydney, E.B. (2011). Screening of microalgae with potential for biodiesel production and nutrient removal from treated domestic sewage. *Applied Energy*, *88*(10), 3291–3294.
Ting, H.H. (2017). Progress in microalgae cultivation photobioreactors and applications in wastewater treatment: A review. *International Journal of Agricultural and Biological Engineering*, *10*(1), 1–29.
Ugwu, C.U. (2008). Photobioreactors for mass cultivation of algae. *Bioresource Technology*, *99*, 4021–4028.
Vassalle, L., Sunyer-Caldú, A., Díaz-Cruz, M., Arashiro, L., Ferrer, I., Garfí, M., and García-Galán, M. (2020). Behavior of UV filters, UV blockers and pharmaceuticals in high rate algal ponds treating urban wastewater. *Water*, *12*, 2658.
Walter Mulbry, S.K.-W. (2008). Treatment of dairy manure effluent using freshwater algae: Algal productivity and recovery of manure nutrients using pilot-scale algal turf scrubbers. *Bioresource Technology*, *99*(17), 8137–8142.
Whitney, S.M. (2011). Advancing our understanding and capacity to engineer nature's CO_2-sequestering enzyme. Rubisco. *Plant Physiology*, *155*, 27–35.
Zeng, X.D. (2012). NaCS-PDMDAAC immobilized autotrophic cultivation of Chlorella sp. for wastewater nitrogen and phosphate removal. *Chemical Engineering Journal*, *187*, 185–192.
Zouhayr Arbib, J.R. (2012). Chlorella stigmatophora for urban wastewater nutrient removal and CO_2 abatement. *International Journal of Phytoremediation*, *14*(7), 714–725.

4 Phycoremediation of Toxic Metals for Industrial Effluent Treatment

Khalida Bloch and Sougata Ghosh
Department of Microbiology, School of Science,
RK University, Rajkot, Gujarat, India

CONTENTS

4.1 INTRODUCTION

Urbanization and industrialization cause surface and subsurface contamination of water bodies by releasing heavy metals, which have become a global concern. Heavy metals are also released into nature through several anthropogenic activities. Heavy metals are nonbiodegradable and have long persistence, which can harm human health and the ecosystem (Salama et al., 2019). As the heavy metals are directly disposed in water bodies, they also have a deleterious effect on the aquatic environment, which limits clean water availability. Several strict regulations are imposed to reduce the concentration of heavy metals in wastewater before discharging into water bodies. Various techniques are used to remove heavy metal from industrial effluents; they include ion exchange, precipitation, extraction by chemicals, hydrolysis, micro-encapsulation, and leaching (Jais et al., 2017). However, these methods have several disadvantages; they are less effective, have a high operating cost, require constant monitoring, are time consuming, depend on parameters such as pH, and require

DOI: 10.1201/9781003165101-4

several chemicals. Hence, bioremediation is an alternate and environmental friendly approach for the effective removal of heavy metals from the environment.

Algae-mediated bioremediation is termed as phycoremediation, which has recently emerged for the removal of heavy metals. Phycoremediation has many advantages. It can be applied in wastewater at very high concentrations, doesn't need biomass synthesis, can recycle and reuse biomass, evidences maximum potential for uptake and removal of heavy metal, doesn't produce toxic chemicals, uses dead biomass, doesn't need a nutrient supply, can be applied in both aerobic and anaerobic conditions, and is cost effective due to its year-round availability. Algae act as prominent agents in the removal of heavy metals (Salama et al., 2019). The mechanisms behind the removal of heavy metal are biosorption, bioreduction, bioaccumulation, and adsorption. This chapter describes how algae removes various heavy metals, such as arsenic, cadmium, chromium, lead, mercury, and selenium. Several factors influencing the efficiency of metal removal are also highlighted.

4.2 HEAVY METAL TOXICITY

In ecological systems, metal plays an integral role whose concentration and availability is maintained by biogeochemical cycles. Various natural process and anthropogenic activities are responsible for releasing metals in the environment. The most metal released into the environment is from automobiles, tanning, mining, petroleum industries, fertilizer use, and manure in agriculture. Metal has been categorized by function and biological effect into essential (such as iron (Fe), manganese (Mn), copper (Cu), zinc (Zn), nickel (Ni)) and nonessential metals (such as mercury (Hg), cadmium (Cd), chromium (Cr), lead (Pb)). Essential metals are required for metabolism because they act as cofactors and coenzymes, and they are tolerated at high concentrations (Ghosh et al., 2020). Nonessential elements do not play any role in biological systems, and they are categorised as heavy metals because they show toxic and adverse effects on ecosystem.

Metals are by nature long lived. Their direct disposal into water bodies and soil leads to an increase in the concentration of heavy metals in the environment. Heavy metal accumulates in microorganisms and aquatic life, hence entering the food chain. These metals exhibit life-threatening damage and are considered as potential pollutants. Biomagnification of Hg in the food chain was the major cause of Minamata disease. Higher concentrations of Pb, Cr, As, and Cr have been found to be involved in causing cancer. They also damage liver, kidney, and nervous systems. Long-term exposure to Ni affects immune systems, blood cells, and reproductive systems. Improper use and disposal of heavy metals into the environment are a major concern due to their affect on the normal functioning of biological systems (Priyadarshini et al., 2019).

4.3 REMOVAL OF HEAVY METALS BY ALGAE

4.3.1 Arsenic

Various anthropogenic activities are considered as a main source of arsenic (As) pollution; these include petroleum refining, non-ferrous smelting, gold mining, and agricultural use of arsenical herbicides and pesticides. Arsenic exists in natural water in two oxidation states: As(III) (arsenite, AsO_3^{2-}) and As(V) (arsenate, AsO_4^{3-}), among which As(III) is 10 to 60 fold more poisonous than As(V) (Tien et al., 2004). Prolonged exposure to As(III) may lead to cancers of kidney, bladder, liver, lungs, and skin and can also cause hypertension, cardiovascular disease, melanosism, hyper pigmentation, and dermal pigment and skin disease. Considering arsenic's severe health hazards, the maximum contaminant level (MCL) of arsenic has been revised and reduced to 10 μg/L from 50 μg/L by the World Health Organization (WHO) in 1993 (WHO, 1993) and the European Commission in 2003 (European Commission Directive, 98/83/EC, 1998) (Ghosh et al., 2021).

Microalgae *Botryococcus braunii* was cultivated in BG11 culture medium ($NaNO_3$ 1.5g/L, K_2HPO_4 0.04 g/L, $MgSO_4.7H_2O$ 0.075 g/L, $CaCl_2.2H_2O$ 0.036g/L, citric acid 0.006 g/L, Na_2EDTA 0.001g/L, Na_2CO_3 0.02 g/L, ferric ammonium citrate 0.006g/L, trace metal mix A5). The algal inoculum was prepared in nutrient agar plates and was allowed to incubate for 7 d at 28°C. The pH of the medium was kept 9.0 with 1 N NaOH and 1 N HCl. Different conc entrations of arsenic-enriched water were used to investigate the removal of arsenic by microalgae *B. braunii*. The algal culture was aseptically withdrawn and was transferred to 100 mL of fresh medium supplemented with arsenic water. A controlled environment was maintained, and illumination was supplied at 2,000 lux by fluorescent lamp. The culture was allowed to incubate for 360 h and 5 mL of culture was taken at regular intervals to analyze the arsenic. The cultures were centrifuged and 10% v/v of HNO_3 was added to dilute the supernatant. The pH had a critical role on the phycoremediation of arsenic, and the percentage (%) removal of both As (III) and As (V) was higher in the basic medium than in acidic medium. Maximum 85.22 and 88.15% removal was observed at pH 9.0 after 144 h. Low amount of phycoremediation was obtained below pH 3. The highest growth of algae was found at pH 9.0. Reduction in pH resulted in the reduction in the algal growth. As pH of the system increased, the surface protonation decreased. As the pH increased above 9.0, the primary amine group present in algal biomass got deprotonated, leading to further reduction in the phycoremediation. The process involves three reasons: 1) development of negatively charged adsorbate due to the accumulation of hydroxyl ions (OH^-) on the surface of algal biomass; 2) ionization of weak acidic group of algal biomass; 3) repulsive force between negatively charged surface of algae and arsenic anions. Effect of inoculum size was also checked on the process of phycoremediation where 2–20% (v/v) range of inoculum was used. No change in As removal was found in with 2–8% (v/v) inoculum, while the maximum removal was obtained in 10% (v/v). With 1% (v/v) inoculums, 71.30% and 80.46% removal was observed in case of As(III) and As(V), respectively. Use of 10% (v/v) inoculums resulted in an enhanced bioremoval up to 85.22% and 88.15% for As(III) and As(V), respectively. The effect of contact time on phycoremediation was checked and maximum removal in As(III) and As(V) was obtained at 144 h, which increased from 40% to 85.22% and 42% to 88.15% respectively. Varying concentrations of arsenic (2,000, 1,800, 1,500, 1,200, 1,000, 800, 500, 200, 100 and 50 mg/L) were used to check the metal ion impact on *B. braunii*. Increases in concentration resulted in increased toxicity. Removal of As(III) and As(V) was 85.22 and 88.15 at 50 mg/L of arsenic, which decreased to 58.25 and 61.21 at 2,000 mg/L of arsenic, respectively. Scanning electron microscope (SEM) analysis was carried out for native microalgae. As(III) and As(V) loaded microalgal biomass showed smooth surface in case of native algae, which converted to rough surfaces after arsenic loading (Podder and Majumder, 2016).

In another study, biotreatment of arsenic-contaminated wastewater was carried out using microalgae *Chlorella pyrenoidosa*. The pure strain of *C. pyrenoidosa* (NCIM no. 2738) was grown on BG 11 medium at 2,000 lux at 28°C. The algal inoculum (10% (v/v)) was transferred into 100 mL of medium supplemented with As(III) and As(V) followed by incubation for 16 d at 28°C under 2,000 lux illumination with photoperiod of 12:12 h dark/light cycle. The highest metal removal was obtained at pH 9.0, while pH below 3 facilitated the removal of As(III) and As (V) up to 13.9% and 15.9%, respectively. The maximum phycoremediation was obtained at pH 9.0, where 80% and 83.5% removal of As(III) and As(V) was evident, respectively. Further, the inoculum density, which indicated the highest phycoremediation at 10% v/v inoculum size in case of both As(III) and As(V). With an increase in As(III) concentration from 50 to 2,000 mg/L, the biomass concentration was reduced from 2.362 g/L to 1.631 g/L, whereas in the case of As(V), it was found to decrease from 2.321g/L to 1.855 g/L, due to the toxicity of arsenic. Arsenic also affected the chlorophyll content in algae where the highest chlorophyll was found in the absence of arsenic and the lowest content was recorded at 2,000 mg/L concentration of arsenic. Removal of As(III) and As(V) was also affected by the contact time of microalgae as the removal of As(III) and As(V), increased from 38.3% to 81.74% and 40.5% to 85.1%, respectively, when the contact time was varied from 4 to 168 h. At 50 mg/L concentration of As(III) and As(V) led to phycoremediation up to 81.7% and 85.1%, respectively. Scanning electron microscope (SEM) analysis revealed that the surface of the native microalgae was smooth, while it was found to be rough in the case of arsenic-loaded microalgae, as seen in Figure 4.1. Several functional groups present in microalgae aided in phycoremediation. Aliphatic C-H, aldehyde, aliphatic acid C=O stretching were speculated to be responsible for the phycoremediation (Podder and Majumder, 2015).

Removal of arsenic from drinking water was carried out using palladium nanoparticles (PdNPs) synthesized using *Chlorella vulgaris*. The PdNPs synthesized

FIGURE 4.1 Scanning electron micrographs (SEM) (1,500×) of (a) native *C. pyrenoidosa* biomass; (b) As(III) loaded biomass; and (c) As(V) loaded biomass. Reprinted with permission from Podder and Majumder, 2015. Phycoremediation of arsenic from wastewaters by *Chlorella pyrenoidosa*. *Groundw Sustain Dev.* 1(1–2), 78–91. Copyright © 2015 Elsevier B.V.

from the algal extract were spherical with particle size of 15 nm. Initially, 1.73 g of sodium arsenite was dissolved in 1 L of deionized water, which was subjected to treatment using phycogenic PdNPs at various retention times, pH, arsenic concentration, and PdNPs concentration. Percent adsorption of arsenic via PdNPs was checked at definite periods using 0.5 g/L arsenic and 1 g/L PdNPs. The arsenic concentration was reduced with higher rate after 5 min. An arsenic elimination up to 89.8% and 92.1% was observed after 5 min and 30 min, respectively. The percent adsorption of arsenic may also be affected by pH. A 99.9% adsorption of arsenic was obtained at pH 3 and 4, while 91.2%, 76.4%, 75.9% and 60.3% adsorption took place at pH 5, 6, 7, and 8, respectively. Different concentrations of PdNPs (0.1, 0.25, 0.5, 1, 2, and 4 g/L) were allowed to react with 0.5 g/L arsenic. At 0.5 g concentration, 100% arsenic elimination was obtained (Arsiya et al., 2017).

Biosorption of arsenic from water was investigated using *Cladophora* sp., which was collected from arsenic-rich local tube well tanks and ponds of Chhattisgarh, Himachal Pradesh, and Jammu, states of India. Different types of algae were grown in arsenic-supplemented medium. Algal biomass was allowed to react with arsenic-enriched water to quantify the biosorption rate. Various biomass (5, 10, 20, 25, and 30 g) of algae was placed for 5, 10, 15 days, respectively, to check the biosorption rate. Different algae, like *Cladophora, Chlorodesmis* (hair algae), and *Chlorella,* were identified. Cladophora was found to be more efficient to grow in the presence of arsenic. The highest growth rate of algae in arsenic-enriched water was 2/3 that in arsenic-free water. Biosorption was observed for 10 d with a maximum of 99.8% uptake. The cell surfaces ruptured in case the algal cells were in contact with arsenic as compared to control. Strong, rough, deep groves and ridges were observed, which may be due to the cross linking of metal and chemical groups present in the cell walls of algae. Almost 0.36% of arsenic was found to be attached with the cell wall of algae (Jasrotia et al., 2014).

Another alga, *Scenedesmus* sp., could alter toxicity and transform and accumulate the arsenic from soil. The algal species was isolated from soil and cultivated on Bold's basal medium (BBM) under continuous illumination (200 μmol/photons m^2/sec) at 25°C ± 2°C under shaking conditions. Arsenic-toxicity analysis was carried out using $NaAsO_2$ and $Na_2HAsO_4.7H_2O$ as sources of As(III) and As(V), respectively. The exponentially grown cell (10^5 cell/mL) was inoculated in water containing arsenic at a concentration from 3.75 to 375 mg/L, followed by incubation at room temperature under illumination and shaking. The specific growth rate of the algae was reduced to 60% at 225 mg/L of As (III). The IC_{50} (50% inhibitory concentration) after 72 h for As(III) and As(V) was 196.5 ± 15.2 and 20.6 ± 3.5 mg/L. The algal cell were incubated with different concentration of arsenic till 8 d, after which total arsenic remained in solution was found to be 50% of the initial total concentration. At 0.75 mg/L, 75% of arsenic was in the form of As(III). But when concentration of phosphate was increased from 2–10 mg/L, reduction of As(V) decreased from 75% to < 25%, indicating the influence of phosphate on uptake and reduction of arsenic. Accumulation of arsenic was more with 761.6 μg/g dry weight as compared to 606.2 μg/g dry weight. The BCF value was found to be 808 and 1,015 for As(III) and As(V), respectively (Bahar et al., 2013).

4.3.2 Cadmium

Cadmium is another highly toxic metal for which the U.S. Environmental Protection Agency (USEPA) and the Indian standard code (IS 10500) have set 2 mg L^{-1} and 1 mg L^{-1}, respectively, as the permissible limits of cadmium in effluent when discharging to a water body. Several algae are reported to remove cadmium most effectively (Shen et al., 2018).

The phycoremediation potential of the green algae *Botryococcus brurauni* was evaluated against several heavy metals in wastewater and synthetic solutions. The freshwater macroalgae was collected and maintained in Fog's medium. Further, the algal culture was enriched in Bold's basal medium. Wastewater was collected from the Yamuna River, followed by sedimentation and filtration to remove solid particles. Metal-contaminated water (25 mL) was inoculated with algae and was kept under incubation for 10 d at 24°C ± 1°C at 3,500–4,000 lux with a light/dark cycle of 16/8 h. The cadmium sulfate ($3CdSO_4.8H_2O$) was used as source of Cd^{+2} that was reacted with the algal culture. After 90 min, 89% removal of cadmium was achieved. After 14 d, more significant change was observed in physiochemical parameters of wastewater, such as pH (8.9), color (pale white), hardness (63mg/L), alkalinity (233 mg/L), total nitrogen (156 mg/L), nitrate (1.9 mg/L), phosphate (1.2 mg/L), chloride (469 mg/L), ammonium nitrogen (17 mg/L), total dissolved solids (TDS) (453 mg/L), chemical oxygen demand (COD) (68 mg/L), biological oxygen demand (BOD) (76 mg/L), and dissolved oxygen (DO) (3.8 mg/L). Chlorophyll content was highly suppressed by cadmium which resulted in toxicity toward algae (Uddin and Lall, 2019).

Freshwater charophytes, particularly green algae *Chara aculeolata* and *Nitella opaca,* were used for the removal of cadmium. Both macroalgae were grown in glass aquaria containing 10% Hoagland's nutrient solution at 25°C ± 2°C under 2,200 lux for 12/12 h light/dark at pH 5.5. After 1 week, the algae were exposed to various concentrations of cadmium (0.25 and 0.5 mg/L) using $Cd(NO_3)_2$ in 10% Hoagland's nutrient solution. Reduction in relative growth rate was observed in Cd (0.5 mg/L). Cadmium caused softening in the thallus due to toxicity, as shown in Figure 4.2. Both the algae showed reduction in total chlorophyll, chlorophyll a and b, and carotenoid at high cadmium concentration. *C. aculeolata* showed 19.9 and 3.8 mg/g total chlorophyll and carotenoid, respectively, when exposed to highest concentration of cadmium. Interestingly, 100% cadmium (with initial concentration of 0.25 mg/L) removal was achieved with *C. aculeolata* whereas *N. opaca* was sensitive to cadmium at 0.5 mg/L concentration. Bioaccumulation studies indicated that metal accumulation in algae increasesd with the increase in concentration of cadmium. In case of *N. opaca* a high bioaccumulation up to 1,544.3 µg/g at 0.5 mg/L initial cadmium concentration (Sooksawat et al., 2013).

Bioremediation of cadmium (Cd(II)) was also reported using pellets derived from water-hyacinth immobilized with *Chlorella* sp. (Shen et al., 2018). Initially, the *Chlorella* sp. (FACHIB-31) was collected and grown on modified basal medium. Water-hyacinth from local river was dried at 105°C for 24 h. The biochar was formed by microwave pyrolysis carried out at 1,100 W for 3.5 min. The pellets of water hyacinth leaf (WLp), water hyacinth root (WRp), and water hyacinth leaf

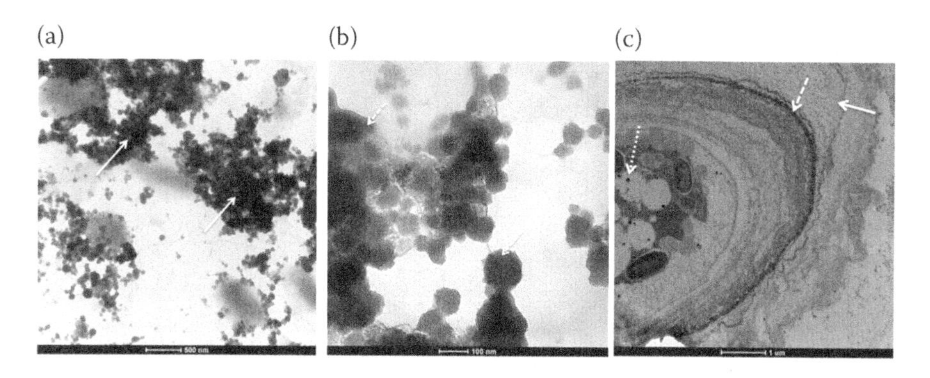

FIGURE 4.2 TEM images of phycosynthesized iron nanoparticle; (a) presence of nanoiron outside the cell (solid arrows); (b) capping of nanoiron by biomolecules (dashed arrows); (c) localization of nanoiron on the cell surface (dashed arrows), inside (dashed arrows) and outside (solid arrows) the cell. Reprinted with permission from Subramaniyam et al., 2015. *Chlorococcum* sp. MM11—A novel phyco-nanofactory for the synthesis of iron nanoparticles. *J Appl Phycol*, 27(5), 1861–1869. Copyright © 2015 Springer Science+Business Media Dordrecht.

biochar (WLBp) were prepared. The biochar was prepared using an ethanol soluble binder such as phenol resin at a ratio of 3:1 under 15 MPa pressure with a diameter of 1.2 cm. The WLp, WRp, WLBp, and WRBp were evaluated with or without immobilized microalgae. All of them were placed in 100 mL of Cd(II) solution at 10 mg/L concentration. The formation of biofilm by *Chlorella* sp. was checked on four carriers. The surface hydrophilicity of WLBp and WRBp was greater than raw biomass carriers. Almost 89.30% immobilization efficiency was obtained on WLBp carrier while 92.45 ± 0.5% removal of Cd(II) was achieved with WLBp immobilized with *Chlorella* sp. Further, *Chlorella* sp. immobilized on WLBp showed 15.51 ± 2.73 mg/g cadmium bioaccumulation efficiency and also remained viable at concentration of 10 mg/L of Cd(II) (Shen et al., 2018).

In another study, palladium oxide nanoparticles synthesized by brown algae *Dictyota indica* extract was used for cadmium removal. *D. indica* collected from the coastal region of Oman was used to prepare the algal powder, which was soaked in water at 60°C for 20 min resulting in the algal extract. Next, 10 mL of 1 mM palladium chloride solution ($PdCl_2$) was reacted with 20 mL of algal extract at 60°C for 2 h on magnetic stirrer. The dark brown color indicated formation of palladium oxide nanoparticles, which had diameters ranging from 8–43 nm with an average diameter of 19 nm. At pH 3, palladium oxide nanoparticles showed the lowest cadmium removal up to 13.8%, which increased to 40.06% at pH 5. At various pH like 7, 8, 9, 10, and 10.5, cadmium removal was 49.56%, 68.62%, 82.82%, 91.07% and 97.81%, respectively. At low pH, protonation is high, but cadmium existd in cationic form; hence, there is a competition between protons and cadmium ions to attach to adsorbent surfaces, causing higher adsorption. The effect of contact time was also investigated on adsorption processes. By increasing the adsorbent contact time, the rate of adsorption increases. At time 0, cadmium removal was only 14.6% which increased to 39.4% after 2 min and 42.6% after 5 min. After 20 min, removal was 50.7% which further increased to 52.1% after 30 min. By increasing the adsorbent concentration, removal rate also increased. With an adsorption concentration of 200 mg/L, cadmium removal efficiency was up to 87.8% which increased up to 96.5% at 900 mg/L adsorbent dosage. Similarly, initial cadmium concentration also played a critical role in determining the removal efficiency. At 0.2 mg/L initial concentration of cadmium, cadmium removal was 97.9% while with a high initial concentration of 10 mg/L, a significantly low cadmium removal (38.22%) was obtained (Shargh et al., 2018).

Algal tissue of *Hydrodictyon reticulatum* was used for the removal of cadmium. The algal tissues were collected, washed, and dried in oven at 40°C. The algal tissues were powdered with grinder and sieved to obtain 120 μm particle sizes. Further, it was washed with distilled water, dried, and treated with 7.5 M NaOH for 24 h under-agitation at room temperature. The treated biomass was reacted with various concentrations of cadmium (5–50 mg/L) to evaluate the extent of bioremoval. Adsorption experiments were carried out at 25°C ± 1°C in solutions containing 20 mg/L of cadmium supplemented with 5 g/L of biomass or alkaline treated *H. reticulatum* biomass. Various groups like hydroxyl, carboxyl, and sulfonate groups on algal cell surface acted as adsorption sites. The removal efficiency and adsorption capacity improved with alkali treatment. Cadmium removal efficiency of untreated bioadsorbent was 59.3% to 76.5% at pH 2 and 3, while alkali-treated bioadsorbent showed 72.3% to 92.8% of cadmium removal at pH 2 and 4, respectively. The highest removal efficiency was 93% with an adsorption capacity of 3.71 mg/g for the alkali-treated biosorbent. Cadmium adsorbed per unit mass of bioadsorbent decreased with the increase in the bioadsorbent dose from 0.10 to 0.3 g/50 mL in case of both untreated and alkali-treated bioadsorbent. On the other hand, the removal efficiency increased with increasing bioadsorbent doses. Maximum adsorption capacity was obtained at pH 6 and with a bioadsorbent dose of 0.10g/50 mL for both untreated and alkaline-treated bioadsorbent (Ammari et al., 2017)

Green algae *Scenedesmus*-24 was used for the biological sequestration of cadmium, as well as retention of adsorbed cadmium sulphide (CdS) nanoparticles. The strain was designated as IMMTCC-24, which was collected, isolated, and cultivated in self-designed photobioreactors; they were supplied with modified Bold's basal medium (Stein, 1973) and allowed to incubate at 30°C ± 2°C for 16:8 light/dark cycles. The cultivation was carried for 10 d by providing cool white fluorescent light at an intensity of 50 μmol photon/m^2/s. The pH was adjusted to 6.8 by passing 99.9% CO_2 gas every 24 h. The algal cells were collected and centrifuged at 2,330 rpm for 20 min and were further resuspended in milli-Q water. The culture (10% v/v) was activated by incubating on orbital shaker at 130 rpm, 30°C under light of 50 μmol photon/m^2/s. Absorbance of the biomass was measured at 680 nm every 24 h. The maximum adsorption of Cd (87%) was obtained at pH 6. Use of the microalgal biomass in concentrations of 0.5 to 1.5 g/L resulted in enhancement of cadmium removal from 87% to 98%. At concentration of 1.5 to 3 g/L, high removal efficiency (99%) was obtained. Cadmium concentration between 10 to 200 mg/L resulted in the decrease in the removal efficiency from 98.4% to 60.5%. But the adsorption of cadmium in algal cell increased from 3.93 to 48.404 mg/g when initial concentrations of cadmium were increased from 10 to 200 mg/L. Maximum biomass of algae was obtained on the ninth day with a specific growth rate of 0.354/d. At 15 mg/L of cadmium, maximum biomass obtained was 2 g/L while at 25 mg/L, it was 1.8 g/L. Functional groups such as carboxylic, amide, amino, hydroxyl, and phosphate aided in metal binding. The size of nanoparticles ranged from 150–175 nm, while the washed biomass elliptical cells unexposed to metal were oval, whereas cells exposed to cadmium possessed irregular cell walls. The cells exposed to metal showed dark particles with a size range from 120 to 175 nm (Jena et al., 2015).

4.3.3 Chromium

Chromium is widely used for chrome plating, petroleum refining, manufacture of pigments, leather tanning, and wood processing, and thus, industries associated with these activities are potential sources of hexavalent chromium [Cr(VI)]. The polluting Cr(VI) oxyanions are highly water soluble, cell-permeable, and transportable in carcinogenic and mutagenic water sources. Further, this hazardous element can also cause liver damage, pulmonary congestion, and skin irritation, leading to ulcer formation. Hence, ISI and WHO have fixed the maximum tolerance limit for total chromium for public water supplies at 0.05 mg/L (Nandi et al., 2017). Various algae are known to remove and convert hazardous Cr(VI) most effectively. Iron nanoparticles were synthesized from *Chlorococcum* sp. MM11 and used in the bioremediation of chromium. *Chlorococcum* sp. M11 was isolated from soil and was allowed to grow on Bold's basal medium at 25°C ± 2°C on shaker under continuous illumination (200 µmol photons/m^2/s light). Synthesis of iron nanoparticles was done using the exponentially grown algal culture. The cells were harvested with the help of centrifugation carried out at 4,000 rpm for 10 min followed by washing thrice with ultrapure water. Algal biomass (0.4 g) was mixed with 5 mL of $FeCl_3$ (0.1 M) and kept under dark conditions on shaker at 120 rpm, 24°C for 48 h. The yellowish-brown color changed from reddish-yellow color after 48 h, indicating formation of nanoiron. The UV-Vis absorbance peak at 293 nm indicated the presence of nanoiron. TEM analysis confirmed that particles were spherical with 20–50 nm size, as shown in Figure 4.2. The nanoparticles were present in the outer membrane of algal cell. Intra- and extra-cellular localization of nanoiron was also observed. The synthesis of nanoparticles was due to the adhesion of iron to the cell surface followed by reduction. The carbohydrates and proteins were involved in the bioreduction and capping process. The nanoiron synthesized from alga reduced Cr (VI) to Cr(III). Nanoiron was reduced 92% of 4 mg/L of Cr(VI). The reduction was dose dependent. The increase in nanoiron concentration increased the reduction of Cr(VI) (Subramaniyam et al., 2015).

4.3.4 Lead

Lead (Pb) is also a highly toxic metal that has severe adverse effects on health and the environment. The potential sources of lead are industries associated with battery manufacturing, printing, pigments, fuels, photographic materials, and explosives. Lead toxicity leads to impairment of the renal, reproductive, and nervous systems. Various algae and their associated products are used for the bioremediation of lead. Hydrochar is solid waste generated from hydrothermal liquefaction (HTL), which leads to pollution. HTL-derived hydrochar prepared from microalgae was used for the removal of Pb(II) from wastewater. Using HTL process, untreated hydrochar (U_char) was generated from wastewater filamentous microalgae. Pretreatment of hydrochar by oil extraction was initiated by the removal of organic compounds with the help of 175 mL of acetone in a Soxhlet extractor at 56°C until the color turned transparent. In CO_2 activation, 0.5 g of hydrochar was heated at 5°C/min ramping at 800°C followed by 45 min incubation under 30 cm^3/min CO_2 flow. Four types of hydrochars were formed: U_char (untreated hydrochar), EN_char (oil extracted), UA_char (CO_2 activated), and EA_char (oil extracted and CO_2 activated). UA_char was also mixed with 5 M NaOH at 60°C and filtered with deionized (DI) water; it was kept in a vacuum overnight for drying, which was named as UAW_char (NaOH-washed CO_2-activated char). Batch adsorption was performed by mixing 100 mg hydrochar with 25 mL of Pb solution (concentrations ranging from 10–100 mg/L) for 5 h at 25°C with continuous stirring. The Ash content in U_char was high as compared to algal biomass. Ash rich U-char had low HHV (higher heating values). The carbon content of U_char was 46.4 wt% while the oxygen/carbon of U_char (0.85) was higher than feedstock (0.75). SEM images in Figure 4.3 revealed that filamentous algal biomass possessed a flat and smooth surface while it had irregular aggregation with many small cracks in case of U_char and UA_char. Absence of methyl, methylene, and aromatic functional groups were noted in UA_char, EN_char and EA_char. Hence, it can be concluded that oil extraction and CO_2 activation may remove these functional groups from the surface of the hydrochar. PH, metal ions, charge on the surface of adsorbent, and degree of ionization played an important role in determining the extent of adsroption. The adsorption increased with the increase in pH (2.0–5.0) but decreased when pH went above 5.0. Maximum adsorption of 2.2 mg/g Pb(II) was obtained at pH 5.0. Metal ion concentration acted as an essential factor in the process of adsorption. Adsorption increased with increase in concentration of Pb(10–30 mg/L), which may be due to the transfer of Pb(II) into the internal structure of hydrochar. At lower concentrations, Pb(II) adsorbed only on the outer surface of the hydrochar, which resulted in decreased adsorption ability. The adsorption capacity of UA_char and EA_char was significantly higher than U_char and EN_char, respectively for Pb(II). There might be more adsorption sites generated during CO_2 activation on hydrochar

FIGURE 4.3 SEM images of (a) algal biomass; (b) U_char; and (c) UA_char. Reprinted with permission from Yu et al., 2021. Waste-to-wealth application of wastewater treatment algae-derived hydrochar for Pb (II) adsorption. *MethodsX*. 8, 101263. Copyright ©2021 The Author(s). Published by Elsevier B.V.

that had attributed to almost 100% removal of Pb(II) by UA_char and EA_char. Further, the UA_char was treated with NaOH to remove silica and was evaluated for adsorption capacity. After treatment with alkali, adsorption of UAW_char increased from 12.83 mg/g to 25.00 mg/g (Yu et al., 2021).

In another study, iron oxide nanoparticles (Fe_3O_4-NPs) were synthesized from brown sea weeds, *Padina pavonica* and *Sargassum acinarium* for bioremediation of lead. The seaweeds were washed and placed at −20°C for 3 d followed by grinding into fine powder. Then 1 g of freeze-dried seaweeds was boiled in 100 mL distilled water under continuous stirring for 15 min. The extract mixed with 0.1 mol/L of ferric chloride ($FeCl_3$) in ratio of 1:1 ratio resulting in immediate formation of Fe_3O_4-NPs. After complete reaction, the nanoparticles were recovered by centrifugation and washed with ethanol, followed by drying under vacuum at 40°C. TEM analysis showed that the Fe_3O_4-NPs were spherical with size ranging from 10 to 19.5 nm for *P. pavonica* and 21.6 to 27.4 nm for *S. acinarium*. Sea weeds are found to be very rich in lipids, minerals, vitamins, polysaccharides, protein, and polyphenols. Hence, these phytochemicals may act as metal-reducing agents, as well as capping agents. Fe_3O_4-NPs alginate beads were prepared by adding 10 mL of 4% w/V sodium alginate into Fe_3O_4-NPs. The mixture was added drop wise into 3.5% calcium chloride solution with continuous stirring. The Fe_3O_4-NPs alginate beads were tested for the removal of lead from aqueous solution. About 88% and 75% lead removal took place at pH 6 after 45 min when *P. pavonica* and *S. acinarium* synthesized nanoparticles were used, respectively. It was observed that Fe_3O_4-NPs alginate beads synthesized from *P. pavonica* and *S. acinarium* were able to remove 91% and 78% of lead, respectively after 75 min. It was observed that the rate of adsorption of lead by adsorbent decreased with decreased adsorption sites and initial concentration of lead (EL-Kassas et al., 2016).

Freshwater charophytes, *Chara aculeolata* and *Nitella opaca,* were used for the removal of lead from the wastewater. Samples of the macroalgae were collected and grown in 20 L of aquaria containing 10% Hoagland's nutrient solution at 25°C under 2,000 lux light for 12/12 light/dark cycle at pH 5.5. Various concentrations of lead were exposed to the algal solution. Reduction was observed in relative growth rate of algae as the concentration of lead increased, indicating that higher concentration caused toxicity. Lead caused the softening of thallus in both the algal species. *N. opaca* was found to be more sensitive to lead, and it showed reduction in total chlorophyll content. When *C. aculeolata* was exposed to 10 mg/L of lead, a decrease in total chlorophyll (13.9 mg/g) and carotenoid (2.5 mg/g) was observed. Removal of lead on day 3 by *C. aculeolata* was 96.8 ± 1.3% at 5 mg/L of lead and was 94.1 ± 0.6% at 10 mg/L of lead. On sixth day removal was 90.6 ± 0.9% at 5 mg/L of lead and 95.0 ± 1.6% at 10 mg/L of lead. Metal removal by *N. opaca* on 3 d was 93.2 ± 1.0% at 5 mg/L of lead and was 88.5 ± 3.0% at 10 mg/L of lead. Similarly, on sixth day, the removal was 76.5 ± 3.2% at 5 mg/L of lead, while it was 43.2 ± 7.7% at 10 mg/L of lead. Bioaccumulation of lead was checked, and it was found to be dependent on the initial concentration of the metal. Interestingly, at low concentration, *N. opaca* accumulated more lead as compared to *C. aculeolata. N. opaca* showed highest accumulation (21,657.0 μg/g) at 10 mg/L of lead. (Sooksawat et al., 2013).

Biosorption of lead was also demonstrated by brown algae (*Sargassum hystrix, Sargassum natans, Padina pavonia*), red algae (*Gracilaria corticata, Gracilaria canaliculata, Polysiphonia violacea*) and green algae (*Ulva lactuca, Cladophora glomerata*). All the algal species were harvested, washed, and dried, followed by grinding. Initially, 100 mg of dried biomass was mixed with 50 mL of lead solution (pH 4.5) at 30°C under shaking conditions. The highest amount of lead removal was exhibited by the brown algae (S. hystrix, S. natans and P. pavonia). Brown algae possess high metallic binding sites due to the presence of polysaccharides, such as alginates, xylofucoglycuronans, xylofucoglucans, and homofucans. The carboxyl and sulfate group present in polysaccharide act as a main metal sequestering sites. Maximum removal of lead by Padina and Sargassum was obtained at pH 4–5, which significantly decreased at pH 1.0. The adsorbed lead could be recovered by simple nondestructive treatments (Jalali et al., 2002).

Other microalgae like *Nannochloropsis* sp. (NN) and *Chlorella vulgaris* (CV) were used for the synthesis of silver (AgNPs) and gold nanoparticles (AuNPs) for removing lead. *Nannochloropsis* sp. and *C. vulgaris* were isolated from wastewater and grown in BG 11 medium. The biomass formed was harvested on 22nd day by centrifuging at 5,000 rom for 20 min. Further, it was washed with distilled water. Algal powder (1 g) was added into 100 mL of distilled water and boiled at 100°C for 20 min, followed by centrifugation to prepare the cell-free extract of microalgae. The AgNPs and AuNPs were synthesized by reacting 10 mL of algal extract with 90 mL of 1 mM of silver nitrate ($AgNO_3$) and 1 mM of gold chloride ($AuCl_3$), respectively. AuNPs synthesized from *C. vulgaris* and *Nannochloropsis* sp. showed purple color change after 2 h with signature peak at 520 nm and 535 nm, respectively, in UV-visible spectra. The CV-AgNPs and NN-AgNPs ranged in size from 27 to 90 nm and 41 to 88 nm, respectively, with spherical shape. The spherical polydispersed CV-AuNPs and NN-AuNPs ranged in size from 7 to 20 nm and 7 to 26 nm, respectively. The AgNPs and AuNPs synthesized from both the algae were used to treat pharmaceutical effluent for 12 h. Adsorption of lead by CV-AgNPs and NN-AgNPs was 66.10% and 68.86%, respectively, while for CV-AuNPs and NN-AuNPs, lead removal was found to be 57.41% and 66.53%, respectively (Adenigba et al., 2020).

4.3.5 Mercury

Among various toxic metals, mercury is in the third position, according to the list of priority pollutants created by the Agency for Toxic Substances and Disease Registry (ATSDR). Mercury has deleterious effects on the ecosystem and human health; various forms of mercury are rapidly absorbed by organisms and slowly eliminated, resulting in transmition, bioaccumulation, and biomagnification along the food chain (Fabre et al., 2021). Several algae are able to remove mercury most effectively. Algal consortium was reported to uptake of methylmercury (MeHg). Periphyton was collected from then shoots of *Scheanoplectus californicus* and cultivated on a freshwater medium under fluorescent light. Further, it was subcultured five times at an interval of 5–6 days. The most abundant alga in the consortium was *Oedogonium* sp., which was further selected for methylmercury uptake. The periphyton's green algae consortium possessed abundant filamentous *Oedogonium* sp., coccoid *Chlorella* sp., and *Scenedesmus* sp. In case of biovolume (μm^3), it was 54% abundance in *Oedogonium* sp, 40% in *Chlorella* sp., and 6% *Scenedesmus* sp. Organic carbon present in 0.02 g/L of algal biomass was 0.1 mM C. Three dosage of algae (0.005, 0.01, and 0.02 g/L) was supplemented with 40 mL of Alga-Gro® freshwater medium containing methylmercury (10 ng MeHg/L). Temperature was maintained at 16.5°C ± 0.5°C with pH 6.8 ± 0.3. The samples were collected at 0.17, 0.3, 0.5, 0.75, 1, 1.5, 2, 3, 6 and 24 h and were filtered with polytetrafluoroethylene (PTFE) filter membrane with pore size 0.22 μm followed by acidification with 1% v/v HCl. Loss of MeHg was 33% after 24 h, while in the first 6 h, the loss of MeHg was negligible. The uptake of MeHg by algae linearly increased with the increase in the initial concentration of MeHg (100 and 180 ng/L). The maximum uptake reached a plateau at 2,863 ± 142 ng MeHg/g due to saturation of adsorption sites on cell membranes of algae. After 6 h, MeHg bound to the cell membrane and inside the cell gave a ration of 0.7, which indicated that 70% MeHg was adsorbed onto the cell wall of algae (Quiroga-Flores et al., 2021).

Several green, brown, and red macroalgae from saline water are also used to remove mercury. Two green (*Ulva intestinalis* and *Ulva lactuca*), brown (*Fucus spiralis* and *Fucus vesiculosus*), and red (*Gracilaria* sp. and *Osmundea pinnatifida*) marine macroalgae were collected and physical and morphological characterization was carried out. The macroalgae were washed and cultured in synthetic sea water at 22°C ± 2°C, followed by interaction with Hg (II) contaminated water. The relative growth rate (RGR) at 72 h was 0.6, 0.5, and 3.1 for *U. intestinalis* at 50 $\mu g/dm^3$, 200 $\mu g/dm^3$ and 500 $\mu g/dm^3$ Hg(II) concentrations, respectively. RGR for *U. lactuca* was 3.2, 0.5 and 3.2 at 50 $\mu g/dm^3$, 200 $\mu g/dm^3$ and 500 $\mu g/dm^3$ Hg(II) concentrations, respectively. After 72 h, maximum Hg(II) removal was 99.9%, 99.6%, and 98.2% for *Gracilaria* sp., *U. lactuca* and *U. intestinalis*, respectively. *U. intestinalis* showed maximum 98.6% and 97.3% removal at 200 and 500 $\mu g/dm^3$ concentration of Hg(II). *F. vesiculosus* showed 80.9% removal of Hg(II) (200 $\mu g/dm^3$) while *O. pinnatifida* showed 69.8% removal at concentration of 500 $\mu g/dm^3$. *U. intestinalis* showed maximum uptake of Hg(II) with a high external contact area, which aided in the removal of cations while *U. lactuca* showed the lowest. At 500 $\mu g/dm^3$, the amount of Hg in macroalgae was 74 μg/g in *F. vesiculosus* and 209 μg/g in *U. intestinalis* (Fabre et al., 2021).

4.3.6 Selenium

Selenium (Se), although an essential element in trace amounts for all vertebrates, can affect adversely impairing the function of metabolic enzymes at higher concentrations. The selenium is mostly present as the selenate oxyanion (SeO_4^{2-}) in oxygenated wastewater, which results in severe ecotoxicity to aquatic life in watersheds (Johansson et al., 2016). Macroalgal biochar can effectively remove selenium, which is mostly attributed to biosorption. Another alga named *Oedogonium* was also used. Fe-biochar was produced from GW and *Oedogonium* by initially soaking the biomass of the algae in $FeCl_3$ solution (1.5%, 4%, 8% and 125% Fe^{+3} w/v) for 24 h at 20°C under shaking (75 rpm) with a density of 25g biomass/L of Fe^{+3} solution. The biomass

TABLE 4.1
Algae-Mediated Removal of Toxic Metals

Metal	Algae	Removal Efficiency (%)	Mechanism	References
Arsenic	*Botryococcus braunii*	88.15	Bioreduction	Podder and Majumder, 2016
	Chlorella pyrenoidosa	85.1	Bioreduction	Podder and Majumder, 2015
	Chlorella vulgaris	99.8	Adsorption	Arsiya et al., 2017
	Cladophora sp.	99.8	Biosorption	Jasrotia et al., 2014
	Scenedesmus sp.	50	Biotransformation and bioaccumulation	Bahar et al., 2013
Cadmium	*Botryococcus brurauni*	89		Uddin and Lall, 2019
	Chara aculeolata	100	Bioaccumulation	Sooksawat et al., 2013
	Chlorella sp.	92.45	Bioaccumulation	Shen et al., 2018
	Dictyota indica	97.9	Adsorption	Shargh et al., 2018
	Hydrodictyon reticulatum	93	Bioadsorption	Ammari et al., 2017
	Nitella opaca	–	Bioreduction	Sooksawat et al., 2013
	Scenedesmus 24	99	Adsorption	Jena et al., 2015
Chromium	*Chlorococcum* sp. MM11	92	Bioreduction	Subramaniyam et al., 2015
Lead	*Chara aculeolata*	95.6 ± 1.6	Bioaccumulation	Sooksawat et al., 2013
	Chlorella vulgaris	66.10	Adsorption	Adenigba et al., 2020
	Microalgae	100	Adsorption	Yu et al., 2021
	Nannochloropsis sp.	68.86	Adsorption	Adenigba et al., 2020
	Nitella opaca	88.5 ± 3.0	Bioaccumulation	Sooksawat et al., 2013
	Padina pavonica	88	Adsorption	El-Kassas Hala et al., 2016
	Padina pavonia	–	Biosorption	Jalali et al., 2002
	Sargassum acinarium	75	Adsorption	El-Kassas Hala et al., 2016
	Sargassum hystrix	–	Biosorption	Jalali et al., 2002
	Sargassum natans	–	Biosorption	Jalali et al., 2002
Mercury	*Chlorella* sp.	–	Absorption	Quiroga-Flores et al., 2021
	Fucus vesiculosus	80.9	–	Fabre et al., 2021
	Gracilaria sp.	99.9	–	Fabre et al., 2021
	Oedogonium sp.	–	Absorption	Quiroga-Flores et al., 2021
	Osmundea pinnatifida	69.8	–	Fabre et al., 2021
	Ulva lactuca	99.6	–	Fabre et al., 2021
	Ulva intestinalis	98.6	–	Fabre et al., 2021
Selenium	*Cladophora hutchinsiae*	96	Biosorption	Tuzen and Sarı, 2010
	Gracilaria sp.	58	Biosorption	Johansson et al., 2016
	Oedogonium sp.	100	Biosorption	Johansson et al., 2016

was then filtered and rinsed with deionized water, followed by drying in an oven for 24 h at 60°C. Eventually, Fe-biochar was formed after slow pyrolysis at 300°C. A higher yield of Fe-biochar was obtained from GW ranging from 67.7% to 80.1%, while *Oedogonium* showed 33.3% to 62.1% yield. It was observed that concentration of Fe on biosorbent depends on the concentration of Fe^{+3} in solution. *Gracilaria* showed lower Fe concentration. GW-based Fe-biochar possessed 25 g/kg of Fe when exposed to 1.5%w/v Fe^{+3} solution, which increased to 50 g/kg Fe when treated with 12.5% of Fe^{+3} solution. *Oedogonium* based Fe-Char possess 59 g/kg and 180 g/kg Fe content at 1.5% and 12.5% w/v Fe^{+3} solution, respectively. The biosorption capacity of the biochar significantly increased on treatment with Fe. The highest biosorbent capacity (q) for both GW and *Oedogonium* Fe-biochar was 6.8 mg/g and 17.03 mg/g, respectively. GW and *Oedogonium* based Fe-biochar showed selenium removal up to 7–58% and 77–100%, respectively (Johansson et al., 2016).

In another study, *Cladophora hutchinsiae* was used for the removal of Se(IV). The green algae were collected, thoroughly washed and dryed at 333 K for 48 h followed by size reduction for the biosorption experiment. Algal biomass was reacted with 1,000 mg/L Se(IV) under shaking conditions. It was speculated that various chemical interactions such as exchange of ions between metal and hydrogen atoms of carboxyl, hydroxyl, and amide groups present in biomass resulted in biosorption of selenium. Maximum Se(IV) sorption (96%) was obtained at pH 5. The biosorption increased with the increase in biomass concentration up to 8g/L while the fastest biosorption was obtained at 60 min of contact time. It was also observed that rate of biosorption decreased from 96% to 60% by increasing the temperature from 20°C to 50°C. The mean free-energy value suggested that biosorption of Se(IV) using algae was due to chemical ion exchange, which was found to be 10.9 kJ/mol (Tuzen and Sarı, 2010). (Table 4.1).

4.4 CONCLUSION AND FUTURE PROSPECTS

Heavy metal interaction with algae causes morphological and physicochemical changes. To overcome the toxic effect, algae have evolved a defense mechanism. Biosorption, bioconversion, and bioaccumulation are the predominant mechanisms that enable the algae to tolerate high concentrations of heavy metals and resist the associated toxicity. The underlying mechanisms behind heavy metal removal by algae have shown the potential role of biomolecules associated with the cell boundary. Mostly alginates, xylofucoglycuronans, xylofucoglucans, and homofucans (apart from lipids, minerals, vitamins, polysaccharides, protein, and polyphenols) are responsible for biosorption (Jalali et al., 2002).

It is important to exploit this rich diversity of algal biomolecules for the synthesis of nanoparticles from the heavy metals. Various biological sources like bacteria, fungi, and medicinal plants are reported to show nanobiotechnological promise for synthesizing gold, silver, copper, platinum, palladium nanoparticles, and zinc nanoparticles (Ghosh et al., 2016a–d; Rokade et al., 2017; Shende et al., 2017; Bhagwat et al., 2018; Ghosh, 2018; Rokade et al., 2018; Shende et al., 2018; Shinde et al., 2018; Jamdade et al., 2019; Robkhob et al., 2020). Hence, the algal biomolecules might play vital role in the reduction of the metal ions to their corresponding nanoparticles followed by their stabilization (Ghosh et al., 2015a–d; Joshi et al., 2019). Further, phycogenic nanoparticles can be functionalized with therapeutic agents, targeting ligands, or contrast agents for biomedical applications (Salunke et al., 2014; Adersh et al., 2015; Kitture et al., 2015; Karmakar et al., 2020). Similarly, various enzymes like azoreductases, laccases, peroxidises can be functionalized on the surface of the phycogenic nanoparticles to bring about dye degradation as well. Likewise, conjugating enzymes, such as dehalogenase, organophosphorus hydrolase, organophosphorus anhydrolase, arbofuran hydrolase, carboxyl esterase, phosphotriesterase, and pyrethroid hydrolase on to the surface of the biogenic nanoparticles would help to degrade the pesticide residues in the wastewater (Sant et al., 2013; Bloch et al., 2021; Ranpariya et al., 2021). In view of the background, use of nanotechnology and synthesis of nanoparticles from algae can act as a novel and potent approach for effective bioremediation of toxic metals, dyes, pesticides, and other refractory pollutants. Rational optimization and scale-up would make phycoremediation one of the most competent complementary and alternative wastewater treatment processes.

REFERENCES

Adenigba, V.O., Omomowo, I.O., Oloke, J.K., Fatukasi, B.A., Odeniyi, M.A., and Adedayo, A.A. (2020). Evaluation of microalgal-based nanoparticles in the adsorption of heavy metals from wastewater. *IOP Conf. Ser.: Mater. Sci. Eng.*, *805*(1), 012030.

Adersh, A., Kulkarni, A.R., Ghosh, S., More, P., Chopade, B.A., and Gandhi, M.N. (2015). Surface defect rich ZnO quantum dots as antioxidant inhibiting α-amylase and α-glucosidase: A potential anti-diabetic nanomedicine. *J. Mater. Chem. B.*, *3*, 4597–4606.

Ammari, T.G., Al-Atiyat, M., Abu-Nameh, E.S., Ghrair, A., and Jaradat, D. (2017). Bioremediation of cadmium contaminated water systems using intact and alkaline-treated alga (*Hydrodictyon reticulatum*); naturally grown in an ecosystem. *Int. J. Phytoremediation*, *19*(5), 453–462.

Arsiya, F., Sayadi, M., and Sobhani, S. (2017). Arsenic (III) adsorption using palladium nanoparticles from aqueous solution. *J. Water Environ. Nanotechnol.*, *2*(3), 166–173.

Bahar, Md. M., Megharaj, M., and Naidu, R. (2013). Toxicity, transformation and accumulation of inorganic arsenic species in a microalga *Scenedesmus* sp. isolated from soil. *J. Appl. Phycol.*, *25*, 913–917.

Bhagwat, T.R., Joshi, K.A., Parihar, V.S., Asok, A., Bellare, J., and Ghosh, S. (2018). Biogenic copper nanoparticles from medicinal plants as novel antidiabetic nanomedicine. *World J. Pharm. Res.*, *7*(4), 183–196.

Bloch, K., Pardesi, K., Satriano, C., and Ghosh, S. (2021). Bacteriogenic platinum nanoparticles for application in nanomedicine. *Front. Chem.*, *9*, 624344.

El-Kassas, H.Y., Aly-Eldeen, M.A., and Gharib, S.M. (2016). Green synthesis of iron oxide (Fe_3O_4) nanoparticles using two selected brown seaweeds: Characterization and application for lead bioremediation. *Acta Oceanol. Sin.*, *35*(8), 89–98.

Fabre, E., Dias, M., Henriques, B., Viana, T., Ferreira, N., Soares, J., Ointo, J., Vale, C., Pinheiro-Torres, J., Silva, C.M., and Pereira, E. (2021). Bioaccumulation processes for mercury removal from saline waters by green, brown and red living marine macroalgae. *Environ. Sci. Pollut. Res. Int.*, 28, 1–12.

Ghosh, S. (2018). Copper and palladium nanostructures: A bacteriogenic approach. *Appl. Microbiol. Biotechnol.*, *101*(18), 7693–7701.

Ghosh, S. (2020). Toxic metal removal using microbial nanotechnology. In Rai, M., and Golinska, P. (Eds.), *Microbial Nanotechnology*. CRC press, Boca Raton.

Ghosh, S., Chacko, M.J., Harke, A.N., Gurav, S.P., Joshi, K.A., Dhepe, A., Kulkarni, A.S., Shinde, V.S., Parihar, V.S., Asok, A., Banerjee, K., Kamble, N., Bellare, J., and Chopade, B.A. (2016c). *Barleria prionitis* leaf mediated synthesis of silver and gold nanocatalysts. *J. Nanomed. Nanotechnol.*, *7*, 4.

Ghosh, S., Gurav, S.P., Harke, A.N., Chacko, M.J., Joshi, K.A., Dhepe, A., Charolkar, C., Shinde, V.S., Kitture, R., Parihar, V.S., Banerjee, K., Kamble, N., Bellare, J., and Chopade, B.A. (2016d). *Dioscorea oppositifolia* mediated synthesis of gold and silver nanoparticles with catalytic activity. *J. Nanomed. Nanotechnol.*, *7*, 5.

Ghosh, S., Harke, A.N., Chacko, M.J., Gurav, S.P., Joshi, K.A., Dhepe, A., Dewle, A., Tomar, G.B., Kitture, R., Parihar, V.S., Banerjee, K., Kamble, N., Bellare, J., and Chopade, B.A. (2016b). *Gloriosa superba* mediated synthesis of silver and gold nanoparticles for anticancer applications. *J. Nanomed. Nanotechnol.*, *7*, 4.

Ghosh, S., Jagtap, S., More, P., Shete, U.J., Maheshwari, N.O., Rao, S.J., Kitture, R., Kale, S., Bellare, J., Patil, S., Pal, J.K., and Chopade, B.A. (2015c). *Dioscorea bulbifera* mediated synthesis of novel $Au_{core}Ag_{shell}$ nanoparticles with potent antibiofilm and antileishmanial activity. *J. Nanomater.*, *2015*, Article ID 562938.

Ghosh, S., More, P., Derle, A., Kitture, R., Kale, T., Gorain, M., Avasthi, A., Markad, P., Kundu, G.C., Kale, S., Dhavale, D.D., Bellare, J., and Chopade, B.A. (2015d). Diosgenin functionalized iron oxide nanoparticles as novel nanomaterial against breast cancer. *J. Nanosci. Nanotechnol.*, *15*(12), 9464–9472.

Ghosh, S., More, P., Nitnavare, R., Jagtap, S., Chippalkatti, R., Derle, A., Kitture, R., Asok, A., Kale, S., Singh, S., Shaikh, M.L., Ramanamurthy, B., Bellare, J., and Chopade, B.A., (2015a). Antidiabetic and antioxidant properties of copper nanoparticles synthesized by medicinal plant *Dioscorea bulbifera*. *J. Nanomed. Nanotechnol.*, *S6*, 007.

Ghosh, S., Nitnavare, R., Dewle, A., Tomar, G.B., Chippalkatti, R., More, P., Kitture, R., Kale, S., Bellare, J., and Chopade, B.A. (2015b). Novel platinum-palladium bimetallic nanoparticles synthesized by *Dioscorea bulbifera*: Anticancer and antioxidant activities. Int. J. Nanomedicine, *10*(1),7477–7490.

Ghosh, S., Patil, S., Chopade, N.B., Luikham, S., Kitture, R., Gurav, D.D., Patil, A.B., Phadatare, S.D., Sontakke, V., Kale, S., Shinde, V., Bellare, J., and Chopade, B.A. (2016a). *Gnidia glauca* leaf and stem extract mediated synthesis of gold nanocatalysts with free radical scavenging potential. *J. Nanomed. Nanotechnol.*, *7*, 358.

Ghosh, S., Selvakumar, G., Ajilda, A.A.K., and Webster, T.J. (2021). Microbial biosorbents for heavy metal removal. In Shah, M.P., Couto, S.R., and Rudra, V.K. (Eds.), *New trends in removal of heavy metals from industrial wastewater* (pp. 213–262). Elsevier B.V., Amsterdam, Netherlands.

Jais, N.M., Mohamed, R., Al-Gheethi, A., and Hashim, M.A. (2017). The dual roles of phycoremediation of wet market wastewater for nutrients and heavy metals removal and microalgae biomass production. *Clean Technol. Environ. Policy*, *19*, 37–52.

Jalali, R., Ghafourian, H., Asef, Y., Davarpanah, S.J., and Sepehr, S. (2002). Removal and recovery of lead using nonliving biomass of marine algae. *J. Hazard. Mater.*, *92*(3), 253–262.

Jamdade, D.A., Rajpali, D., Joshi, K.A., Kitture, R., Kulkarni, A.S., Shinde, V.S., Bellare, J., Babiya, K.R., and Ghosh, S. (2019). *Gnidia glauca* and *Plumbago zeylanica* mediated synthesis ofnovel copper nanoparticles as promising antidiabetic agents. *Adv. Pharmacol. Sci.*, *2019*, 9080279.

Jasrotia, S., Kansal, A., and Kishore, V.V.N. (2014). Arsenic phyco-remediation by *Cladophora* algae and measurement of arsenic speciation and location of active absorption site using electron microscopy. *Microchem J.*, *114*, 197–202.

Jena, J., Pradhan, N., Aishvarya, V., Nayak, R.R., Dash, B.P., Sukla, L.B., Pnada, P.K., and Mishra, B.K. (2015). Biological sequestration and retention of cadmium as CdS nanoparticles by the microalga *Scenedesmus*-24. *J. Appl. Phycol.*, *27*(6), 2251–2260.

Johansson, C.L., Paul, N.A., de Nys, R., and Roberts, D.A. (2016). Simultaneous biosorption of selenium, arsenic and molybdenum with modified algal-based biochars. *J. Environ. Manage*, *165*, 117–123.

Joshi, K.A., Ghosh, S., and Dhepe, A. (2019). Green synthesis of antimicrobial nanosilver using *in-vitro* cultured *Dioscorea bulbifera*. *Asian J. Org. Med. Chem.*, *4*(4), 222–227.

Karmakar, S., Ghosh, S., and Kumbhakar, P. (2020). Enhanced sunlight driven photocatalytic and antibacterial activity of flower-like ZnO@MoS_2 nanocomposite. *J. Nanopart. Res.*, *22*, 11.

Kitture, R., Chordiya, K., Gaware, S., Ghosh, S., More, P.A., Kulkarni, P., Chopade, B.A., and Kale, S.N. (2015). ZnO nanoparticles-red sandalwood conjugate: A promising anti-diabetic agent. *J. Nanosci. Nanotechnol.*, *15*, 4046–4051.

Nandi, R., Laskar, S., and Saha, B. (2017). Surfactant-promoted enhancement in bioremediation of hexavalent chromium to trivalent chromium by naturally occurring wall algae. *Res. Chem. Intermed.*, *43*, 1619–1634.

Podder, M.S., and Majumder, C.B. (2015). Phycoremediation of arsenic from wastewaters by *Chlorella pyrenoidosa*. *Groundw. Sustain. Dev.*, *1*(1–2), 78–91.

Podder, M.S., and Majumder, C.B. (2016). The use of artificial neural network for modelling of phycoremediation of toxic elements As (III) and As (V) from wastewater using *Botryococcus braunii*. *Spectrochim. Acta A Mol. Biomol. Spectrosc.*, *155*, 130–145.

Priyadarshini, E., Priyadarshini, S.S., and Pradhan, N. (2019). Heavy metal resistance in algae and its application for metal nanoparticle synthesis. *Appl. Microbiol. Biotechnol.*, *7*,1–20.

Quiroga-Flores, R., Guedron, S., and Acha, D. (2021). High methylmercury uptake by green algae in Lake Titicaca: Potential implications for remediation. *Ecotoxicol. Environ. Saf.*, *207*, 111256.

Ranpariya, B., Salunke, G., Karmakar, S., Babiya, K., Sutar, S., Kadoo, N., and Kumbhakar, P., Ghosh, S. (2021) Antimicrobial synergy of silver-platinum nanohybrids with antibiotics. *Front. Microbiol.*, *11*, 610968.

Robkhob, P., Ghosh, S., Bellare, J., Jamdade, D., Tang, I.M., and Thongmee, S. (2020). Effect of silver doping on antidiabetic and antioxidant potential of ZnO nanorods. *J. Trace Elem. Med. Biol.*, *58*, 126448.

Rokade, S., Joshi, K., Mahajan, K., Patil, S., Tomar, G., Dubal, D., Parihar, V.S., Kitture, R., Bellare, J.R., and Ghosh, S. (2018). *Gloriosa superba* mediated synthesis of platinum and palladium nanoparticles for induction of apoptosis in breast cancer. *Bioinorg. Chem. Appl.*, *2018*, 4924186.

Rokade, S.S., Joshi, K.A., Mahajan, K., Tomar, G., Dubal, D.S., Parihar, V.S., Kitture, R., Bellare, J., and Ghosh, S. (2017). Novel anticancer platinum and palladium nanoparticles from *Barleria prionitis*. *Glob. J. Nanomedicine*, *2*(5), 555600.

Salama, E.S., Roh, H.S., Dev, S., Khan, M.A., Abou-Shanab, R.A.I., Chang, S.W., and Jeon, B.H. (2019). Algae as a green technology for heavy metals removal from various wastewater. *World. J. Microbiol. Biotechnol.*, *35*, 75.

Salunke, G.R., Ghosh, S.,Santosh, R.J., Khade, S., Vashisth, P., Kale, T., Chopade, S., Pruthi, V., Kundu, G., Bellare, J.R., and Chopade, B.A. (2014) Rapid efficient synthesis and characterization of AgNPs, AuNPs and AgAuNPs from a medicinal plant, *Plumbago zeylanica* and their application in biofilm control. *Int. J. Nanomedicine*, *9*, 2635–2653.

Sant, D.G., Gujarathi, T.R., Harne, S.R., Ghosh, S., Kitture, R., Kale, S., Chopade, B.A., and Pardesi, K.R. (2013). *Adiantum philippense* L. frond assisted rapid green synthesis of gold and silver nanoparticles. *J. Nanoparticles*, *2013*, 1–9.

Shargh A.Y., Sayadi, M., and Heidari, A. (2018). Green biosynthesis of palladium oxide nanoparticles using *Dictyota indica* seaweed and its application for adsorption. *J. Water Environ. Nanotechnol.*, *3*(4), 337–347.

Shen, Y., Zhu, W., Li, H., Ho, S.-H., Chen, J., Xie, Y., and Shi, X. (2018). Enhancing cadmium bioremediation by a complex of water-hyacinth derived pellets immobilized with *Chlorella* sp. *Bioresour. Technol.*, *257*, 157–163.

Shende, S., Joshi, K.A., Kulkarni, A.S., Charolkar, C., Shinde, V.S., Parihar, V.S., Kitture, R., Banerjee, K., Kamble, N., Bellare, J., and Ghosh, S. (2018). *Platanus orientalis* leaf mediated rapid synthesis of catalytic gold and silver nanoparticles. *J. Nanomed. Nanotechnol.*, *9*, 2.

Shende, S., Joshi, K.A., Kulkarni, A.S., Shinde, V.S., Parihar, V.S., Kitture, R., Banerjee, K., Kamble, N., Bellare, J., and Ghosh, S. (2017). *Litchi chinensis* peel: A novel source for synthesis of gold and silver nanocatalysts. *Glob. J. Nanomedicine*, *3*(1), 555603.

Shinde, S.S., Joshi, K.A., Patil, S., Singh, S., Kitture, R., Bellare, J., and Ghosh, S. (2018). Green synthesis of silver nanoparticles using *Gnidia glauca* and computational evaluation of synergistic potential with antimicrobial drugs. *World J. Pharm. Res.*, *7*(4), 156–171.

Sooksawat, N., Meetam, M., Kruatrachue, M., Pokethitiyook, P., and Nathalang, K. (2013). Phytoremediation potential of charophytes: Bioaccumulation and toxicity studies of cadmium, lead and zinc. *J Environ Sci.*, *25*(3), 596–604.

Stein, J.R. (1973). *Handbook of phycological methods: Culture methods and growth measurements*. Cambridge University Press, Cambridge. 448.

Subramaniyam, V., Subashchandrabose, S.R., Thavamani, P., Megharaj, M., Chen, Z., and Naidu, R. (2015). *Chlorococcum* sp. MM11—A novel phyco-nanofactory for the synthesis of iron nanoparticles. *J. Appl. Phycol.*, *27*(5), 1861–1869.

Tien, V.N., Chaudhary, D.S., Ngo, H.H., and Vigneswaran, S. (2004). Arsenic in water: Concerns and treatment technologies. *J. Ind. Eng. Chem.*, *10*, 337–348.

Tuzen, M., and Sarı, A. (2010). Biosorption of selenium from aqueous solution by green algae (*Cladophora hutchinsiae*) biomass: Equilibrium, thermodynamic and kinetic studies. *Chem. Eng. J.*, *158*(2), 200–206.

Uddin, A., and Lall, A.M. (2019). Phycoremediation of heavy metals by *Botryococus brurauni* from wastewater. *Biosci. Biotech. Res. Asia.*, *16*(1), 129–133.

Yu, J., Tang, T., Cheng, F., Huang, D., Martin, J.L., Brewer, C.E., Grimm, R.L., Zhou, M., and Luo, H. (2021). Waste-to-wealth application of wastewater treatment algae-derived hydrochar for Pb (II) adsorption. *MethodsX.*, *8*, 101263.

5 Algal Biomass Production Coupled to Wastewater Treatment

Sanjay Kumar
Department of Chemical Engineering, Faculty of Technology, Marwadi University, Rajkot, Gujarat, India

Arvind Singh
Department of Chemical Engineering & Biochemical Engineering, Rajiv Gandhi Institute of Petroleum Technology Jais, Amethi, Uttar Pradesh, India

Rupak Kishor
Department of Chemical Engineering, Maulana Azad National Institute of Technology, Bhopal, Madhya Pradesh, India

CONTENTS

DOI: 10.1201/9781003165101-5

5.1 INTRODUCTION

Global freshwater consumption has grown by a factor of six over the last century and has been growing at a pace of about 1% per year since the 1980s (as per United Nations World Water Development report). In contrast, the rate of growth in freshwater use has slowed in most Organization for Economic Cooperation and Development (OECD) member countries, where per capita water-use rates are among the highest in the world. It continues to rise in the majority of emerging economies, as well as middle- and lower-income countries. A large portion of this expansion may be ascribed to population increase, economic development, and altering consumer patterns. In addition, pollution of surface water has become a major environmental concern worldwide, as has the current problem of water shortage due to contamination of huge freshwater bodies that render the water unfit for human use. This contamination is a major problem in developing nations where heavy metal pollution, eutrophication, persistent organic pollutants, sewage, and acidity endanger human health.

Wastewater is viewed as a combination of liquid or water-carried waste evacuated from homes, institutions, commercial and industrial enterprises, as well as groundwater, surface water, and storm water. Municipal wastewater comprises a significant concentration of oxygen-depleted wastes, pathogenic or disease-causing agents, and organic substances. Industrial wastewater, on the other hand, contains heavy metals, dyes, medicinal chemicals, and agrochemical compounds. It may also include hazardous substances.

Traditional wastewater remediation employs a variety of physical, chemical, and biological methods to remove particles, organic debris, and, in some cases, nutrients from wastewater. Preliminary treatment, primary treatment, secondary treatment, and tertiary and/or advanced wastewater treatment are general phrases that define distinct levels of treatment.

The goal of preliminary treatment is to remove coarse particles and other big items commonly found in raw wastewater. The physical techniques of sedimentation and flotation are used in primary treatment to remove organic and inorganic materials. During primary treatment, about 25%–50% of the entering bio-chemical oxygen demand (BODs), 50%–70% of the total suspended solids (SS), and 65% of the oil and grease are removed. Clarifiers or settling tanks are used in the first treatment to remove settleable organics and settleable inorganic particles from wastewater. Secondary treatment is the process of treating wastewater after initial treatment to remove residual organics and suspended particles. The secondary treatment method involves the biological treatment of wastewater in a controlled environment using a variety of microorganisms. Secondary treatment is carried out using various aerobic microorganisms. Based on the use of treated water, tertiary and/or advanced wastewater treatment processes, such as membrane treatment technology, reverse osmosis (RO), electrodialysis (ED), ion exchange (IX), and advanced oxidation processes (AOPs), are utilized.

Recent years have seen a surge of interest in the integration of algal biomass production with secondary wastewater treatment. Algae have the unique advantage to carry out photosynthesis and live in aquatic environments like rivers, oceans, lakes, ponds, and streams. It also can live in freshwater, brackish water, soil, and marine. Algae can be categorized in many types based on color, size, and functions like Rhodophyta (red algae), Chlorophyta (green algae), Paeophyta (brown algae), Xanthophyta (yellow-green alga), Pyrrophyta (fire algae), Chrysophyta (golden-brown algae), and Euglenophyta (euglenoids) reported in (Wang et al., 2016).

In past years, numerous studies on the treatment of various industry-based wastewaters using microalgae have been reported (Salama et al., 2017). Notably, photobioreactors and wastewater treatment plants (such as open oxidation ponds) can be used to produce microalgal biomass while simultaneously treating wastewater. Oxidation ponds are the most common wastewater treatment plants for municipal and industrial wastewater treatment combined with microalgal biomass production.

Algal biomass has more potential to produce different metabolites for many applications; see Table 5.1. The production of algae from wastewater has a low cost; it can be used in many applications, such as biodiesel, bioethanol, and biogas. Treated water also can be used in other applications. The cultivation of microalgae can be done in open and closed pond systems. Closed systems are expensive compared to open pond systems because of design and technical problems; for instance, a single unit cannot be extended 100 cm^2 due to gas-exchange limitations (Do et al., 2019). The usual pH range is 7 to 9, whereas 8 is the optimum for growing algae strains in ponds or lakes. Although water's pH value may be reduced by bubbling carbon dioxide in the culture medium, it should not be too high or too low. Microalgae can sustain temperature ranges between 16°C and 35°C. Outside this range, the cell growth rate may slow. Through photosynthesis, processed microalgae converts inorganic carbon into organic matter; its application is shown in Figure 5.1.

The basic objective of wastewater treatment is to minimise the level of COD and other organic nutrient components in the effluent. Several challenges frequently occur in microalgae-based wastewater treatment systems as compared to the traditional activated-sludge process (ASP) wastewater treatment: (1) Algal treatment methods have a lower COD loading (F/M), whereas ASP may treat wastewaters with a wider COD range. Algae, for example, may grow successfully only in wastewater with low COD (less than 5,000 mg L1). (2) During the growth of algal

TABLE 5.1
Lipid, Protein, and Carbohydrate Content in Microalgae

Microalgae Strain	Lipid (%)	Protein (%)	Carbohydrate (%)	References
Scenedesmus sp.	20.8	–	22.1	Ji et al. (2015)
Desmodesmus sp.	23.51	39.66	24.41	Do et al. (2019)
Phormidium sp.	11	62	16	Cañizares-Villanueva et al. (1995)
Namochlorpsis oculata	36	49.8	7.3	Parsy et al. (2020)
Micratinium reisseri	20	–	–	Ji et al. (2015)
Spirulina sp.	9.09	60.34	17.4	Phang et al. (2000)
Tetraselmis suecica	25.06	50.2	10.62	Andreotti et al. (2020)
Chlamydomonas Sp.	11	51.9	19.5	Behl et al. (2020)
Porphyridium cruentum	0.5–0.8	28–35	17–22	Cecal et al. (2012)
Chlorella sp.	14–22	51–58	12–17	Becker (1994)
Euglena sp.	14–20	39–61	14–18	Becker (1994)
Prymnesium sp.	22–38	28–45	25–33	Ricketts (1966)
Anabaena sp.	4–7	48	25–30	Becker (1994)
Ulva lactuca	4.36	8.44	35.27	Chakraborty and Santra
Catenella repns	5.29	8.42	28.96	(2008)
Lola capillaries	4.05	40.87	22.32	

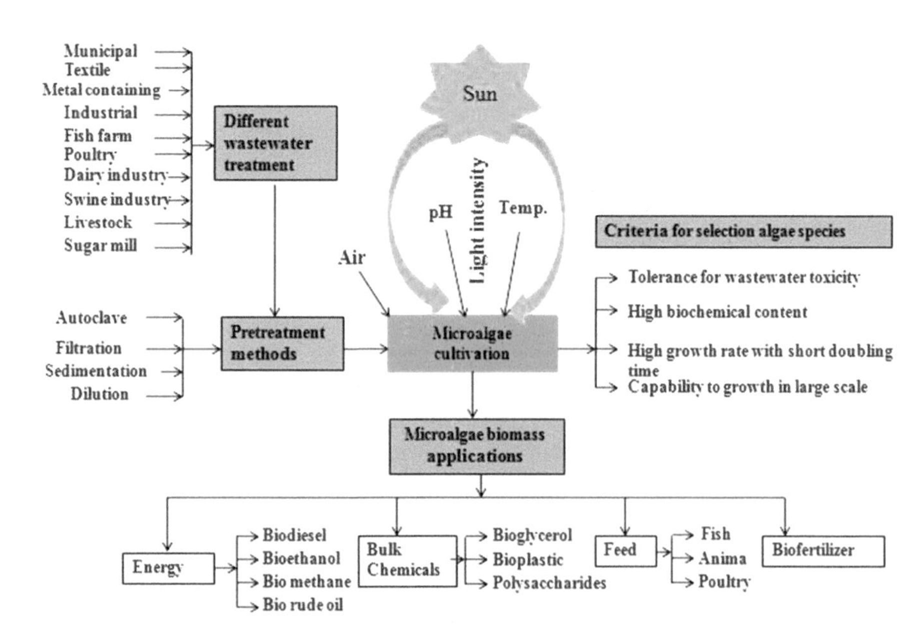

FIGURE 5.1 Different application of microalgae, and types of wastewater-treatment processing using microalgae.

biomass, various organic materials were generated, which increased the COD of the wastewater. To avoid the formation of excess extracellular organic matter, a portion of the microalgal biomass should be collected on time. Most microalgae take longer to collect massive volumes of lipid or carbohydrate (i.e. utilised for biofuel generation), greatly raising the likelihood of contamination and ecological risks. But, at the other side, the organic stuff generated by algae primarily occurred as polysaccharides, which may serve as a dietary supplement for living organisms.

The establishment of algae biomass technology might be an appropriate method for overcoming the aforementioned challenges connected with the microalgae-based wastewater treatment process. As a result, this chapter goes into great depth about microalgae farming. Furthermore, it disucsses the employment of microalgae in dye and metal ion removal, followed by wastewater treatment. Finally, we address the problems of combining wastewater treatment with algal biomass production.

5.2 ADVANTAGES OF USING WASTEWATER AS A CULTURE MEDIUM FOR PRODUCTION OF MICROALGAE

Microalgae production in wastewater has many advantages, such as nutrient recovery in the form of biomass, pollutant and pathogen reduction, reduction of CO_2 emission, and biomass production. The total production of microalgae requires low energy, and therefore, the process is cost effective. Nitrogen and phosphorous in the form of nutrients are essential elements for microalgae growth. The use of commercial fertilizer would increase in cost production and use of agro-industrial wastewater, which is rich in nutrients and a valuable alternative to producing algal biomass (Christenson and Sims, 2011). The removal efficiency of nitrogen and phosphorous from wastewater is directly related to higher biomass productivity. Effluent from the chemical industry has some amount of ammonium, phosphorous, nitrates, and many trace elements as pollutants, which need to be removed to treat wastewater; conventional chemical treatment processes are costlier. Microalgae-bacteria consortia can potentially remove such pollutants while producing biomass. Various microorganisms, such as Escherichia coli from domestic wastewater, are potentially harmful for mankind, and they can be removed by using microalgae. Dissolved oxygen (DO) and pH value of water shows pathogen presence. Effective photosynthetic activity of microalgae results an increase of DO and pH in water. High DO concentration results in damage to pathogen cells, and therefore, its removal. High temperature and pH also results in reduced pathogen from domestic wastewater.

Burning of fossil fuel produces large amounts of CO_2. This CO_2 is released to the environment by many industries, leading to global warming effects. Several techniques are used to study CO_2 capture, such as absorption, adsorption, membrane gas separation, geological sequestration, and bioprocessing from photosynthetic organism (Chisti, 2007). Using microalgae, CO_2 reduction by photosynthesis has advantages because of the higher growth rate of microalgae without any additional supply of nutrients in agro-industrial wastewaters. Throughout the photosynthesis process, microalgae capture large amounts of CO_2 from the environment to produce organic molecules and release O_2 to the atmosphere. So, this is one of the effective ways to capture CO_2 and release O_2 as well as produce biomass. Microalgae has the potential to remove heavy metals like Cr, Cd, Hg, Cu, Ni, and some other metals discharged from industrial wastewater mainly from mining industries, agriculture activity, and pharmaceutical activity. Various removal mechanisms of heavy metal from wastewater have been reported so far, such as ion exchange, adsorption, and covalent bonding or heavy meal precipitation. Those metals get bioaccumulated in cell vacuoles by the diffusion process. In algae, peptides bind with heavy metal and from organometallic complex, which diffuse into vacuoles to control the cytoplasmic concentration of metals (González et al., 2017). Many textile industries discharge effluents directly into water bodies, and, due to that ground water, they become contaminated. Several methods, including chemical, physical, and biological, are used to remove color from wastewater. Algae cultivation is one of the cost-effective ways to remove dye from effluent water. U. lactuca is one of marine algae largely used to remove dye effluent. Due to high surface area and binding affinity in algae, results increase biosorption capacity on the cell surface. Algae have many functional groups, such as carboxyl, hydroxyl, amino, and phosphate, due to which strong attractive force occurs between dyes and the cell wall (Charumathi and Das, 2012). The recent technology of biomass production and its application are discussed in the next section.

5.3 RECENT TECHNOLOGY OF ALGAE BIOMASS PRODUCTION

In recent technology, microalgae biomass production takes place in open ponds, closed photobioreactors, and hybrid production systems. It can grow on sewage and wastewater where nutrients are easily available as nitrogen and phosphorus. Many types of algae are available, but very few of them are best suited for wastewater treatment. Chlorella vulgaris and Chlorella pyrenoidosa in the Chlorella group are widely used; other groups, such as Scenedesmus, are also available for the treatment of wastewater. Different techniques are available for harvesting of microalgae; they inlcude flotation, gravity sedimentation, flocculation process, membrane filtration, and ultrasonic separation process. Microalgae have many applications in different areas, such as energy production, chemical production, animal feed production, and as biofertilizer. This section will give brief ideas about selecting microalgae for wastewater treatment, recent technology available for cultivation of microalgae, harvesting, and its application in various sectors.

5.3.1 Microalgae Species Selection for Wastewater Treatment

More than 3,000 microalgae strains have been identified, as reported in Sheehan et al. (1998), but very few of them are useful for wastewater treatment. Selection of suitable microalgae for wastewater treatment could depend on (i) wastewater characteristics, (ii) overall cost, (iii) energy requirements, (iv) treatment efficiency, and (v) its application. The most preferred characteristics of algae depend on its growth rate, high lipid content, and high tolerance of feasible pollutants (toxic compound, metal ions, CO_2 sinking capacity, and environmental conditions). Those criteria are the main limiting factors for algal growth for a pollutant and nutrient removal efficiency from a wastewater source. For wastewater treatment coupled with biomass production, several microalgae strains are used, such as Chlamydomonas sp., Chlorella sp., Neochloris sp., Scenedesmus sp., Desmodesmus sp., Cosmarium sp. and Nitzschia sp. (Ji et al., 2014; Xiong et al., 2016). Among them, Chlorella sp. are broadly used to remove nitrogen, phosphorous, and other pollutants from wastewater. Scenedesmus sp. are cultivated in piggery, swine, municipal wastewater, and oil mill effluent (Wang, 2016).

Nannochloropsis gaditana marine algae grow using nutrients available in sea-water and clean municipal wastewater. The biomass productivity of marine sea algae contains 0.4 g L^{-1} d^{-1} and lipid accumulation of 20%–25% DW. Laminaria hyperborea and Pelvetia canaliculata brown microalgae strain are capable of removing transition metal ions (Cechinel et al., 2016). Table 5.2 shows few microalgae used in various wastewater treatments.

TABLE 5.2
Various Microalgae Grow in Different Wastewater Treatment Purpose

Industrial Wastewater	Microalgae Strain	Cultivation Period	Nutrient Removal (%)	Biomass Production (g/L)	References
Municipal wastewater	Chlorella kessleri	5	84	2.54	Li et al. (2011)
	Scenedesmus sp.,	4	67	0.99	Zhou et al. (2011)
	Auxenochlorella protothecoides	12	69	2.5	Hu et al. (2012)
	Chlamydomonas reinhardtii	31	–	2	Kong et al. (2010)
Piggery wastewater	Chlamydomonas mexicana	20	63	0.92	Abou-Shanab et al. (2013)
	Scenedesmus obliquus	20	61	0.77	
	Chlorella vulgaris	20	52	0.71	
	Botryococcus braunii	7	–	1.80	An J-Y., et al. (2003)
Industrial wastewater	Chlamydomonas sp.	10	72	1.5	Wu et al. (2012)
Domestic sewage	Botryococcus braunii	14	80	0.64	Sydney et al. (2010)
	Chlorella vulgaris	14	74	0.48	
Urine wastewater	Chlorella sorokiniana	90	–	14.80	Tuantet et al. (2014)

Algae growth rate: Algae specific growth rate (μ) can be determine by the following equation

$$\mu = \frac{lnC_t - lnC_0}{t} \tag{5.1}$$

where t = time in days, C_t is the concentration of biomass (g/L), and C_0 is the initial concentration of biomass (g/L). The specific growth rate of algae calculated using slop of plot lnC_t vs. t (Li et al., 2011).

Lipid content measurement: After cultivation, microalgae were separated by centrifugation at 3,000 rpm for 5 min and then washed multiple times with distilled water; they were dried at 60°C for 12 h and then biomass was reduced further using a hand mill to make it pellet size. Different methods used to measure lipid content in biomass inlcude: (i) Gravimetric quantification methods, (ii) staining methods, (iii) colorimetric SPV method, (iv) TD-NMR method, (v) TLC/HPLC method, and (vi) NIR and FTIR spectra by (Han et al., 2011).

5.3.2 Production of Microalgae

5.3.2.1 Open Pond

Open pond cultivation of microalgae is most widely used for the commercial production of biomass; it is less expensive and can be operated at bulk scale. Microalgae strain of genera *Spirulina, Dunaliella,* and *Chlorella* are used for large-scale cultivation in an open pond system as they are highly adaptive to extreme environmental conditions (Fernández et al., 2013). *Spirulina* can be cultivated in highly alkaline mediums, whereas *Chlorella* in high nutrient concentrations and *Dunaliella* in high salinity water. Growing Spirulina algae requires water and sources of nitrogen, carbon, potassium phosphorus iron, and other trace elements. It converts nutrients into cellular matter, and during photosynthesis, releases oxygen to the environment. *Spirulina* contains high protein (60–70 g/100 g) and good source of minerals, vitamins, c-phycocyanin γ-linolenic acid, and phenols (AlFadhly et al., 2022). Many companies produce *Spirulina* in open ponds, such as Olson company (Bazil), Earthrise Nutritionals (California, USA), Hainan DIC Microalgae (Hainan, China) and Cyanotech Co. (Hawaii, USA) (Uebel et al., 2019). *Chlorella* microalgae produce carbohydrate, lipid, and protein, and it has several advantages, like high biomass yield. It absorbs nutrients from wastewater to grow and treat industrial wastewater, and it can also treat dye effluent from textile waste. It has also several health benefits, such as high protein content 50%–60%, Vitamin B12, iron and vitamin C, fiber and Omega 3 (Katiyar et al., 2021). *Chlorella* is cultivated worldwide by many companies in open pond systems. Taiwan is the largest producer (Wang et al., 2016). *Dunaliella* microalgae produce significant amount of lipids, proteins, and glycerol. It can be used in the desalination of saline water and refinery effluent treatment and heavy metal from wastewater. Raceway photobioreactors are also used for microalgal culture systems; at the commercial scale, they consist of a circuit of parallel channels in which microalgal circulation is promoted by paddle wheels. At a commercial scale, raceway photobioreactors are typically operated at depths of 15–30 cm and with 15–30 cm/s flow rates. The energy consumption of open systems is relatively lower than close systems. Other types of photobioreactors used are circular pond systems with pivoted agitators at the center and inclined cascade systems in which suspension flows from top to bottom and is pumped back to the top, creating high turbulence. In open pond cultivation, it is tough to uphold monoculture due to the contamination of native algae and algae grazers. So monoculture special strategies are used, such as high pH, saline etc.

Factors That Influence Open Pond Reactors

1. **Temperature:** Temperature is the most influential factor responsible for microalgal growth. The optimum temperature varies for different species; however, the optimum temperature for most microalgae ranges from 20°C to 30°C (de Godos et al., 2017). Cultivation of microalgae within this temperature range will give higher nutrient removal efficiency as well as higher biomass production. Outside this temperature range, metabolic rates of microalgae diminish, leading to a slow microalgal growth rate; thus, less biomass productivity and less nutrient removal efficiency results (Molinuevo-Salces et al., 2016). For open pond systems, temperature is not a controllable factor as it is generally dependent on sunlight and environment; therefore, it is essential to select algae as per its environment. Higher temperatures require a supply of cooling water within the reactor and regulation of air temperature by refrigerated air condition; implementation of these two strategies can control overheating temperature.
2. **Light availability:** A reactor should be designed in such a way that the distribution of light should be uniform within the reactor. Microalgae growth rate and biomass production is directly related to sunlight intensity. The photosynthetic activity ranges from 100 to 200 $\mu E/m^2$ day for better microalgal growth, and beyond this range, the growth rate is inhibited by higher radiation (Acien et al., 2017). At higher intensity, cells in deeper layers may be photolimited, and the cells at the surface may be photoinhibited; so, proper light availability is required for cell growth.
3. **pH:** During day, pH rises gradually in microalgae culture due to the consumpsion of inorganic carbon by microalgae. At higher pH, CO_2 absorption capacity decreases and therefore drastically limits algal activity. Several methods can control pH levels in tanks, such as CO_2 injection, acid/base adjustment, and buffer addition (Moheimani, 2013).
4. **Mixing:** Mixing is the most important factor for open pond as it helps to increase mass transfer, turbulence, and light exposure to lower part of the pond and avoid cell accumulation and sedimentation, and concentration and temperature gradient. As mixing can cause cell disruption to some microalgae species, the use of microalgae species should be considered.
5. **Gas transfer and CO_2 delivery:** Microalgal production is a result of photosynthesis, i.e. oxygen is released and carbon dioxide will be consumed. The oxygen transfer rate from water to atmosphere is low, which results in the accumulation of dissolved oxygen, makes the photosynthesis

process very slow, and causes photo-oxidation. The uptake of CO_2 from the atmosphere is low for good cultivation of microalgae, so to meet the carbon requirement, carbon sources are added to the culture.

5.3.2.2 Closed Photobioreactor

In the last few decades, a great variety of closed photobioreactors have been developed. Closed photobioreactors allow greater photosynthetic efficiency, biomass concentration, and productivity (Wang et al., 2012). The advantages of closed systems are the minimization of water evaporation and the reduction of contaminating species, as well as more efficient cell growth. The major limiting factors for commercialization of closed systems are high capital, operational, and maintenance costs. Sunlight plays an important role for this system, where maximum energy is drawn from light for the growth of algae. An optimal pH of 7–9 and a temperature slightly higher (10°C–30°C) than ambient temperature are best suitable for closed systems. A verity of microalgae strains like *Spirulina platensis, Cyanobacteria microalgae, Consortium of microalgae, and S. quadricauda* are generally used for cultivation in closed photobioreactors. Designs of closed photobioreactors should maintain light capturing, uniform distribution of light, proper utilization of light, CO_2/O_2 balance, temperature, pH, mixing, species control, and cleanability of the tank (Wang et al., 2012).

Mainly two different types of closed photobioreactors are designed for algae production and discussed below:

1. **Tubular photobioreactor:** This is one of the most popular types of photobioreactor used for algae production. Straight, spiral, and bent tubes generally are used in different patterns, including different orientations, like horizontal, vertical, and inclined arrangements. For high biomass productivity, generally 0.1 m or less tube diameter is used in all the arrangements. A horizontal tube provides more surface area as compared to vertical tubes because of the decrease in tube design and better angle of inclination to allow light harvesting. For large tube designs, temperature control is the main problem; thus, heat exchangers are generally used to maintain the temperature. Another way to decrease the temperature is to submerge the entire solar tube under water, spray water in the tubes when temperature exceeds the ideal, and shade tubes with dark sheets (Wang et al., 2012; Solovchenko et al. (2014) used distillery wastewater and remarkably removed more than 97% nitrates and 77% phosphates from 50 L alcohol using *Chlorella sorokiniana*. (Salas et al., 2013) used cylindrical reactor of 70 L and grown *Scenedesmus obliquus* species with a biomass productivity of 15.25 $g/m^2/d$. In this type of reactor, contamination is higher than in closed tank systems and lesser than open tank reactors because of the unavailability of proper mixing and lighting conditions.
2. **Flat panel photobioreactor:** This photobioreactor is a flat transparent vessel made of glass or transparent material. The panels used are less than 5–6 cm thickness as light needs to penetrate them for culture. Air bubbles are introduced at the bottom of the tank through perforated tubes for the

better mixing of water. These types of reactors are usually positioned vertically, and plates are mounted at optimized angles to get higher photosynthetic efficiency.

The main advantages for using these types of reactors are (1) high degree of solar light on the surface of the plate, (2) lowered accumulation of dissolved oxygen, and (3) convenience of modular design for scale-up. The demerits of this type of reactor are (1) nonability to control the temperature, (2) formation of algal biofilm, and (3) hydrodynamic stress that impact strains of algae.

5.3.3 Harvesting of Microalgae

Common methods of harvesting microalgae biomass include: coagulation, flocculation (or sedimentation), flotation, centrifugation, membrane filtration, and ultrasonic separation. By the process of coagulation and flocculation, dispersed microalgae cells are efficiently concentrated to form larger aggregates by the addition of polymers or salts. For the settling of aggregates, centrifugation is used; but due to high cost and desired solid concentration gravity settling or dissolved air, floatation is used. Then the aggregates are removed by skimming or flotation. Membrane filtration was initially used for removal of microalgae cells; but due to biofouling, nowadays, ultrasonic separation and electro-coagulation-flocculation are used for microalgae harvesting. Through a membrane filtration process, microalgae cell could recover 100% as compared to other methods of harvesting, and this method can be used for most microalgae strain cases. Another potential biomass-separation technique is tangential-flow-filtration (TFF), in which culture is pumped though the TFF module, where portions of water pass through the membrane and biomass collects on the membrane. In this method, generally backwash is used at some stage in the TFF operation to clean the membrane as well as recover biomass. A few cells of microalgae couldn't stay afloat inside the growth medium due to their mobile length and or the pH value of the medium. In both cases, cells go under the surface to the bottom. Consequently, the sedimentation method could be carried out to split microalgae biomass as a less expensive and variable harvesting system for massive scale operations. In the flotation techniques of harvesting microalgae, generally, coagulants are added like microscopic bubbles at the bottom of the flotation tank in which the microalgae cells get hitched by microscopic bubbles and weaken further cells to flow up and get concentrated. In a electrocoagulation harvesting process, microbubbles of oxygen and hydrogen gas as coagulants are formed at an electrode surface during the process and get attached by microalgae cells and float to the top.

5.3.4 Use of Microalgae

The cultivation of microalgae in wastewater has many applications in different areas, such as energy production, chemical production, animal feed production, and biofertilizer use.

1. **Energy production:** Considerable research and several experiments were done to reach a commercial level of production of biodiesel from lipid-rich microalgae using wastewater. After harvesting, microalgae need to be dried for efficient extraction of lipids. Using anaerobic digestion process of microalgae, biomass could be cost and energy effective. The maximum yield of biogas from microalgae biomass was reported as 200–600 mL/g organic content. Another method to be used with dried microalgae biomass to produce biodiesel is the pyrolysis process. Other applications using microalgae recycle the nutrients in the waste stream. The by-product of biodiesel production from the microalgae lipid is glycerol.
2. **Chemical production:** This is the application of microalgae biomass to the production of bioplastic. The cyanobacterial strains could produce intracellular polyhydroxyalkanotes (PHA) similar to properties like plastic. So, cultivation of cyanobacterial strains in large-scale wastewater could produce bioplastic. Another application of microalgae is ethanol production using fermentation of algal polysaccharides, a sugar, starch, and cellulose. In specific conditions, the carbohydrate can reach up to 70%. Microalgae and cyanobacteria produce exopolysaccharides (EPS) and can be used as a thickener, gelling agent, biolubricants, and anti-inflammatory.
3. **Animal feed production:** Most microalgae have a natural excessive protein content at the same time as having a high oil content, even though manipulation of cultivation occurs. The nutritional properties of numerous microalgae biomass might be on par or exceeding many feed substances. Therefore, the inclusion of microalgae biomass in feed might beautify the quality of the feed. The cost of cultivating microalgae in nutrient-rich wastewater would be considerably less, as there would be little to no requirement for extra nutrients. Spirulina sp. biomass produced in the anaerobically digested sago wastewater at high rate algal pond became suitable for animal feed. Microalgae biomass, produced within the aquaculture effluent, may be used as a component of fish feed.
4. **Human food production:** Over the last decade, microalgae is being used by some countries to make food supplements for human use as tablets, capsules, and liquids. Microalgae is rich in nutrient-like fibers, protein, enzymes, and large quantities of vitamins are present, such as calcium, potassium, iron, iodine, magnesium, A, B1, B2, B6, and C (Priyadarshani and Rath, 2012). However, vitamins cell content will depend on water conditions, harvesting treatment, and biomass drying methods. By using the spirulina (Arthrosphira) strain, Earthrise Nutrritionals (California, USA) makes products like tablets, powder. Cyanotech Crops (Hawaii, USA) and Myanmar Spirulina factory make products in the form of tablet, pasta, chips, and liquid products 3,000 t/year. Other strains of microalgae, like chlorella, dunaliella salina, and aphanizomenon flos-aquae, are used for make products for human use (Pluz and Gross, 2004).

5.4 ALGAL BIOMASS PRODUCTION COUPLED TO WASTEWATER TREATMENT

Large scale microalgae biomass production generally cultivates in open pond systems, and it could produce up to 0.5 g/L. Microalgae biomass composition of nitrogen and phosphorus can be 5.4%–8.7% and 0.7%–1.1%, and from wastewater, it can remove 43.5 and 5.5 mg/L of nitrogen and phosphorous and adsorb pollutants from wastewater. This will help remove other contaminants also. The dissolve organic in municipal wastewater is usually low for the micro organism for the consumption of nitrogen and phosphorus. Microalgae cultivation in municipal wastewater is usually to remove heavy metal, phosphorous, nitrogen, BOD, and pathogens. The cultivation of microalgae in oil and gas industries wastewater is even possible in toxic petroleum compounds and treated water is produced, but some algae strains, like Scenedesmus sp., Dunaliella sp., Chlorella sp., etc. can bear such wastewater (Wang et al., 2016). A number of researches going on for large-scale treatment and cultivation of algae strain of such wastewater by using other types of microalgae strain. Wastewater form poultry, cattle, piggery, and other livestock industries is rich in phosphorous and nitrogen; microalgae are cultivating in this environment, as well as treating this type of industrial wastewater. Nitrogen and phosphorus are the two major nutrients present in the wastewater. They help the production of microalgae, which provides a suitable environment, such as pH, dissolved oxygen, and the presence of ions, to erode the toxicity of the effluent. Microalgae has been proven to be the source of multiple bio-based products, is used as a biofuel, and can also generate freshwater for reuse so they are economically feasible. The major elements in wastewater responsible for this microalgal growth are as follows.

5.4.1 Carbon (C)

Carbon and its compounds from natural as well as industrial sources can be effectively used for the production of microalgae and via photosynthetic activity; it absorbs the toxic components of the effluents. One of the main naturally occurring compounds is carbon dioxide and carbonium ion, which is necessary for the algal pH sustainability, effective biomass production, lipid, hydrocarbon production, and the fatty acid profile of microalgae.

5.4.2 Nitrogen (N)

The major element for the growth of biological macromolecules (including proteins, peptides enzymes, energy transfer fragments, chlorophylls, and genetic constituents) is nitrogen. It can be easily obtained from wastewaters. Eutrophic algae are effective in transforming inorganic nitrogen (such as nitrite, nitrate, nitric acid, ammonium, nitrogen gas, and ammonia) to organic formulas by assimilation. Reduction of inorganic nitrogen leads to the production of more effective ions for the growth of microorganisms, such as chlorella vulgaris in wastewater.

5.4.3 Phosphorus (P)

The basic life molecules of microalgae include DNA, RNA, ATP, proteins, lipids, carbohydrates, and cell membrane materials that require phosphorus for growth and metabolic activities. Wastewater includes phosphorus compounds and ions for photosynthetic activities. During metabolism, P is preferably in the form of ions that combine with organic compounds during phosphorylation; this includes ATP from adenosine diphosphate in the form of energy. Some microalgal species especially use phosphorus originated in organic esters for their growth, including inorganic forms. Eutrophication water is highly enriched with minerals and nutrients, which induce excessive growth of algae and form by superfluous phosphorus for bacterial degradation.

5.4.4 Trace Elements

Microalgae growth also requires trace elements (such as Mn, Zn, Cu, Ca and Fe), which plays an important role in the activation of various cellular enzymes. Manganese (Mn) is a vital nutrient for microorganism biological functions. Manganese catalyses oxygen evolution in the photosynthesis process. Low Mn levels leads to inadequate growth of microalgae. Iron is a vital element of the photosynthetic and respiratory chains. The biochemical significance of Zn has various enzyme and acid synthesis capabilities. The specific growth rate and productivity of cells decreased sharply as concentrations of zinc are increased.

5.5 BIOLOGICAL TREATMENT OF DIFFERENT WASTEWATER COUPLED WITH MICROALGAL PRODUCTION

The composition of wastewater discharged from industrial facilities is complex. Carbon is deficient, but nitrogen and phosphorus are two main components in industrial wastewater, which are capable of supporting algae growth. Utilizing the following properties of wastewater, several authors producted microalgae using different wastewater and main mechanisms involving treatment of different wastewater, as shown in Figure 5.2; details are discussed below.

5.5.1 Production of Algae Biomass with Dye Removal

Textile wastewater contains 10% to 30% of total industrial wastewater in many countries. Wastewater from textiles contains mainly dye, salt, heavy metal (Zn, Cu, As, Cr, Fe etc.), binders, and reducing agents, as well as nitrogen and phosphorous compound. Phycoremediation techniques of dye removal from the wastewater (bioremediation using microalgae) are the most widely used method because of its versatility and capacity compared to bacteria and fungi. Microalgae are the indicators for assessing water quality and ecotoxicity of pollutants. Bioremediation of textile wastewater is a most promising technology due to their potential for simultaneous bioremediation and carbon dioxide mitigation followed by biodiesel production. Microalgae are considered as an important source of decolorizing the

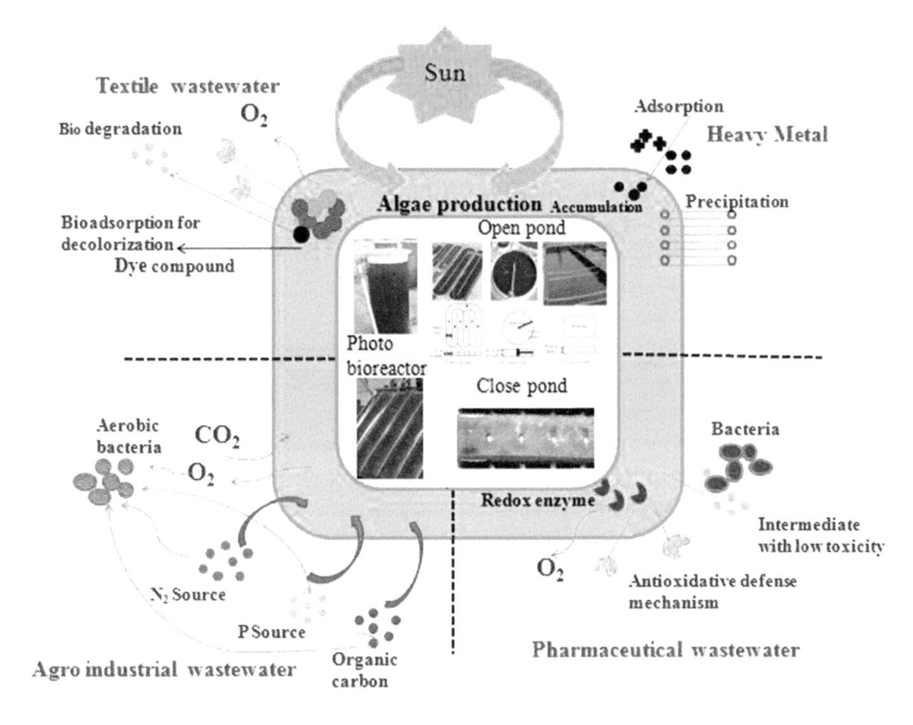

FIGURE 5.2 Mechanisms of various industrial wastewater treatments using microalgae systems.

textile dye effluents. N. elepsosporum recorded the highest percentage decolorisation, followed by C. vulgaris. C. vulgaris has high tolerance to azo dyes and can remove 63%–69% of the color. Caulerpa lentillifera can remove basic dyes, and Synechocystis sp. and Phormidium sp. effectively remove reactive.

Microalgae are photosynthetic creatures found in practically over the globe and in every type of environment. It is often found in both fresh and saltwater and is being widely researched as a bio-sorbent (Mehta and Gaur, 2005). It possesses the bio-sorption ability and ionic force of attraction toward pollutants in wastewater due to its vast surface area and binding ability. Because algal biomass might be regarded as an alternative to standard adsorbent materials for the treatment of HMs and dyes. (a) Microalgae may be cultivated in a variety of environmental circumstances; (b) fast growth rates due to their short cell cycle time; (c) requirement of minimal nutrient concentrations in comparison to other biomass species; (d) no need of agricultural area for cultivation; (e) algae cultivation may be accomplished in wastewater owing to decreased water requirements; and (f) additional uses such as biofuel generation.

Considerable research has been conducted on wastewater treatment by microalgae (Li, et al., 2020; Rani, et al., 2021), and the results show that microalgae might be utilized to treat dye effluents and wastewater containing heavy metal ions. This section highlights the microalgae decolorization of main dye classes, as well as heavy metal removal. In addition, microalgae-based remediation is a more efficient way to wastewater treatment since it can treat wastewater in a single step, as opposed to traditional wastewater treatment, which requires many steps to correct the carbon, nitrogen, and phosphorus ratios (C:N:P) (Rani et al., 2021). It is also an environmentally beneficial solution since it has the ability to transform carbon dioxide into chemical substances and fuel products without generating pollution; therefore, it contributes to the reduction of greenhouse gas emissions. To offset the production expense, microalgae biomass recovered from wastewater treatment may be transformed into useful bio-based goods such as health supplements, bio-hydrocarbons, bio-H_2, and bio-alcohols.

In the subject of wastewater decolorization, the use of microalgae has recently received more attention. Although several documented traditional approaches have shortcomings, biological approaches are now recognised as specific, less energy demanding, effective, and ecologically safe since the partial or total bioconversion of organic contaminants are stable and harmless end products (Bhatia et al., 2017).

Because of its ubiquity and the ease with which it is available in both fresh and sea water, the use of algae for color treatment has attracted a lot of attention (Bhatia et al., 2017). Further, it is a promising candidate for dye bioremediation because it is unaffected during the contact with discharge of harmful azo dyes and may thrive in industrial discharge. The following are the major advantages of systems of wastewater treatment that utilize algae over other biological treatment approaches (Sarkar and Dey, 2021):

- Fast-growing rate aids in maximised-cell population in a short cultivating duration.
- It has a capability for bio-mitigation and CO_2 sequestration.
- Its rate of growth is unaffected by harsh weather conditions.
- There is no rivalry for agricultural land with food crops.
- It is capable of growing in sea, fresh, saline, or wastewater.
- Dead-algae are as efficient as living algal with lower operating and maintenance expenses since it doesn't need nutrients for growth and respiration.
- Improved bioremediation ability and capacity to produce useful by-products.

Microalgae may degrade a variety of dyes, as shown in Table 5.3, with the decrease appearing to be tied to the molecular structure of the dyes as well as the type of algae utilized. The removal of dye by algae is caused by three essentially independent modes of action, the first of which is bio-sorption.

The attaching of contaminants on the surface of microorganisms is referred to as biosorption. Two types of biosorption are: physical and chemical. It also known as physicsorption and chemisorption, respectively.

However, chemisorbed molecules, dye molecules that have been physisorbed, could be eluted. The benefits of biosorption include lower costs, an easy-to-manage technique that does not generate sludge, and the ability to treat huge quantities of wastewater with low pollutant concentrations. Cell wall composition influences algal biosorption, which is important in electrostatic attraction. The presence of

TABLE 5.3

Presents a Summary of Some Studies on Biodegradation of Dyes and Metals by Algae

Dye Removal							
Algae Species	**Dye Removal**	**C_o ***	**pH**	**Temp. (°C)**	**Time (h)**	**Removal Efficiency (%)**	**References**
Chlorella vulgaris	Aniline blue	25	8	–		58	Arteaga, et al. (2018)
	Congo red	50	–	35	216	100	Naji and Salman (2019)
Chara sp.	Malachite green	9.7	6.8	–		57.81	Khataee et al. (2010)
Ulothrix sp.	Methylene blue	90	7.9	37	0.5	86.1	Doğar et al. (2010)
Turbinaria conoides	Acid Blue 9	100	1	33	3.75	87.64	Rajeshkannan et al. (2010)
Padina sanctae-crucis	Methyl violet	10	8	25	1.33	98.85	Mahini et al. (2018)
Spirogyra sp.	Synazol	–	3	30	18	85	Khalaf (2008)
Chlorella sp.	Malachite green	–	10	25	2.5	80.7	Daneshvar et al. (2007)
Ulva lactuca	Methylene blue	100	6	30	1	78	Pratiwi et al. (2019)
	Methylene green	25	8	–	2.83	91.92	Khalaf (2008)
Spirulina platensis	Acid black 210	125	2	60	1	98.55	Al Hamadi et al. (2017)
	Acid blue 7	125	2	60	1.25	97.05	Al Hamadi et al. (2017)
	Reactive black 5	200	5	40	240	80	Ishchi and Sibi (2020)
Chlorella vulgaris	Direct blue 71	200	8	40	240	78	
	Disperse red 1	300	8	40	240	84	
	Malachite green	5	–	–	120	93	Gelebo et al. (2020)
Oscillatoria sp.	Methylene blue	5	–	–	120	66	
	Safranin	5	–	–	120	52	

(Continued)

TABLE 5.3 (Continued)
Presents a Summary of Some Studies on Biodegradation of Dyes and Metals by Algae

Metal Removal						
Algal Strains	**Metals**	**C_o ***	**pH**	**Temp. (°C)**	**Removal Efficiency (%)**	**Reference**
Scenedesmus sp.	Cr	10	0.5–5	–	92.89	Pradhan et al. (2019)
Chlorella vulgaris		227	0.5–5	24	50.7–80.3	Sibi (2016)
Spirulina maxima	Cu	56.6	7.96	26–28	94.9	Chan et al. (2013)
Scenedesmus obliquus		60	5.5	30	72.4–91.7	Li et al. (2018)
Chlorella vulgaris		56.6	7.9	26–28	96.3	Chan et al. (2013)
Scenedesmus sp.	Cd	0.5	6.2–6.5	25	73	Travieso et al. (1999)
Chlorella vulgaris		5	6.2–6.5	25	66	
Chlorella sp.		100	7.4	–	33–41	Wong et al. (2000)
Psuedochlorococcum typicum	Hg	10	7	20	97	Shanab et al. (2012)
Synechocystis sp.	Zn	30	6 –7.2	–	40	Chong et al. (2000)
Scenedesmus sp.		30	6–7.2	–	98	
Chlorella sp.		1–50	8.1–8.6	28	60–70	Kumar and Goyal (2010)
Scenedesmus sp.	Ni	30	6–7.2	–	97	Chong et al. (2000)
Chlorella vulgaris		10–40	7.4	–	33–41	Wong et al. (2000)

Note
* C_o = Initial dye conc. (mg/L).

surface functional groups of an microalgae cell like carboxylate, phosphate, amino, and hydroxyl are accountable for the secretion of numerous pollutants from effluents. Second, biocoagulation is the phenomenon by which dye-molecules coagulate on extracellular biopolymers surface generated by algae when dyes are converted by the metabolic-assisted phenomena (Sarkar and Dey, 2021). These extrinsic long-chain biopolymers have good coagulation characteristics and are composed of surface functional groups. The molecules of dyes present in the aqueous solution are kept adsorb on polymers and settled down. The third method is bioconversion or biodegradation; some Oscillatoria and Chlorella groups are able to degrade azo-dyes, resulting in the creation of aromatic amines, which are then converted to sole organic molecules like CO_2 (Sarkar and Dey, 2021). Factors affecting algal bioremediation processes are:

5.5.1.1 The Effect of pH

pH is one of the most important characteristics influencing biosorption since it impacts solubility of dyes. Moreover, it also affects the states of ionisation of different organic groups, such as amine, carboxyl, and hydroxyl present on surface of adsorbent. It has an impact not just on site dissociation but also on the dyes' solution chemistry (Dotto et al., 2012). Response of any bio-assisted method and reaction become sluggish with changing the optimize value of pH.

5.5.1.2 Influence of Dye Structure

A linear structure with a positively charged and far molecular weight clearly enhances dye adsorption on the algal surface. Various researches have also underlined the impact of heterogeneity. Molecular structures of dye molecules, as well as their ionic radius, affect the biomass surface of fungal and algal biomass during the biosorption process. Other variables include adsorbent particle size, algal biomass type, adsorbent dose/algal concentration, temperature, and beginning dye concentration.

Furthermore, several chemical treatments on algae, such as esterification, amine methylation, formaldehyde treatment, and treatment with acid and cationic surfactants, are being used to investigate the impact of various functional groups present on the surface of microalgae on dye remediation. These treatments have an inhibitory effect on the binding sites, as well as a dispersal of surface ions on the surface of microalgae, which aids in the adsorption of negative charges dye molecules. Moreover, they proposed the following adsorption mechanisms and reasons for enhanced dye uptake: a) creation of complexes with functional groups present on the surface of algae-like amino, carboxyl, phosphate and hydroxyl, b) the process of exchanging ion between molecules of dye and cation, c) Van der Waal's interaction forces brought alkyle chains into contact with dye molecules.

5.5.2 Production of Algae Biomass with Metal Removal

Production of microalgae biomass in different wastewater coupled with metal removal successfully developed in past few decades. Wastewater containing heavy metals discharge from many industries such as textile industries, oil industries, chemical industries, and other industries. The metal substance can't be easily removal by filtration techniques; this requires additional chemical treatment prior to filtration. Using microalgae can remove heavy metal-like Cr, Fe, Zn, Co, etc. efficiently proven by many researchers (Mehta and Gaur, 2005) also grow algae in this type of wastewater by consuming nutrient from waster.

Microalgae can adsorb metal ions by two major processes, one of which is connected to metal ions coordination, physical adsorption, complexation, ion-exchange, and micro-precipitation. This process (bioaccumulation) is devoid of physiological variables and comprises solely the structure and surface group of microalgae cells. The other kind of metal adsorption (biosorption) is connected to algal metabolism, which involves the storage of metal ions inside the microalgae body. Redox reactions, surface complexation, and ion exchange are the key components of biosorption. Microalgae strains have also exhibited remarkable effectiveness in the removal of heavy metals.

Heavy metal absorption by microalgae includes both passive and active biosorption by dead biomass and live microalgae cells. Metal ions in the negatively charged state are physically adsorbed onto the algal cell surface, which includes functional groups such as sulfhydryl (–SH), carboxyl (–COOH), hydroxyl (–OH), and amino ($-NH_2$). Metal ions are translocated through the cell membrane into the cytoplasm throughout active biosorption by Chai et al. (2021)). Algae intracellular polyphosphate bodies can also provide a storage inflow and seclude about all heavy metals (Mehta and Gaur, 2005).

Physical-adsorption, chemisorption, reduction, chelation, and complex formation are all methods for binding extracellular metals. It has been observed that cyanobacterial microalgae create extracellular polymeric substances (EPS) that develop a wrap on the algal cell surface and are capable of converting into liberated polysaccharides (RPS) that disperse in the environment. Ionizable functional groups in EPS and RPS, including phosphoric, hydroxyl, carboxyl, and amine groups, help in the extracellular sorption of metal ions (Pereira et al. (2011)). Biosorption, or the use of nonliving items, was discovered to be more quick and efficient than bioaccumulation (the use of live cells) (Mehta and Gaur, 2005).

According to reported literature, polysaccharides have a significant role in microalgae-assisted sorption of heavy metal. Polysaccharide cells contain the carboxyl group. Sulfonate and amino acids have a little impact on metal sorption. At lower pH (>2), the thiol group is significant in the sorption of metals such as Cd. Despite numerous possible functional groups identified in algae, they do not have any involvement in metal biosorption. Several functional groups present in the surface may be prevented from attaching to metal by steric effect, conformational alterations, or crosslinking, which together alter from surrounding circumstances (pH, competing cations, strength, ionic, or ligands) (Mehta and Gaur, 2005). Surface charge, for example, can be mitigated by interacting with cations or by cross-linking between surface groups having opposite charge.

Heavy metal sorption and removal by microalgae biosorbents are significantly influenced by the initial metal concentration. Metal sorption rises with increasing metal concentration in solution and gets even more saturated after a specific metal concentration. The surface of a microalgae cell has variety of functional groups, each having a different affinity for an ionic species. At elevated metal ion concentration, affinity of functional groups is lower during bioassisted metal ion remediation and vice versa.

Several researchers have indicated that metal ion sorption in batch and continuous systems is affected by the pH of the solution, as shown in Table 5.3. As a result, attempts have been made to determine the optimal pH value in algae-assisted process to achieve maximum efficiency of metal removal. Because of the nature of groups present in microalgae, those responsible for binding metal are acidic, such as carboxyl. Therefore, accessibility of these groups is highly dependent on the pH. At acidic pH, these groups create a surface having a negative charge. These surfaces interact with metallic cationic. This phenomenon is responsible for metal biosorption of heavy metals. At severely acidic pH (2), algal sorption of metals is typically reduced.

For metal binding, wastewaters may contain inorganic ligands such as HCO_3^-, SO_4^{2-}, Cl^-, HS^-, CO_3^{2-}, and organic molecules like amino, acitic, oxialic acids and others. In addition, metal ions present in normal water might be bonded to bio-sorbants and many types of solid surfaces. Metal ion concentration is regulated by complex interaction involving metal ions, ligands, and other ions. The pH of the solution, as well as the degree of hydrolysis and the stability constant of metal-ligand complexes, all have an effect on it.

In general, concentration of free metal ions present in the solution is inversely proportional the pH value. A lower pH value indicates a favorable situation. Furthermore, a greater ratio of free-metal-ion to total metal concentration must be maintained for efficient heavy metal removal by an algal biosorbent. Moreover, metal sorption by living algae is highly influenced by the amount of biosorbent, the presence of anions and cations, temperature, nutritional content, growth rate, and lighting (Mehta and Gaur, 2005).

5.5.3 Municipal Wastewater

In the past, the natural treatment process in lakes and rivers was ample to satisfy the basic requirements. But, due to a large expansion in populations as well as pollutants, there is a high requirement of how to treat wastewater effectively. The outlet pipes from the industries containing toxic compound materials cannot be redirected for sewage treatment without any pre treatment. So, municipal wastewater treatment enters the picture. The treatment involves following steps: First, screening is done to remove large chunks of particles. Then it is passed through a grit chamber to remove sand and small, loose particles of stone, and then sedimentation is done. All these comes under primary treatment. Now there is additional need to remove biological pollutants that play a major role in making water inappropriate for further use. This can be done under secondary and tertiary treatment processes. In this process, the primary function of an aeration tank is to supply oxygen to the tank to encourage the breakdown of organic materials present in wastewater and also to grow bacteria, which further helps in the breakdown of organic materials. After that, it is again redirected to clarify for the settling of heavy material at the bottom. The sludge from the clarifier tank mainly consists of bacteria, which can be used to increase bacterials concentration in the aeration tank. The clear water from the secondary clarifier has much reduced organic materials and is approaching expected effluent specifications, but this process is very costly and energy consuming. So, there is a need of an economizing technique for the secondary and tertiary treatments. The cultivation of microalgae can be used as an effective way to ingest nutrients and deliver aerobic bacteria with the oxygen they need through photosynthesis. For using algae as the source of oxygen and bacteria, various researches were conducted on the growth of algal biomass in the wastewater.

(Arbib et al., 2013) presented nitrogen, phosphorous variation of the wastewater showed no significant differences in terms of specific growth rate and the proper N:P ratio for achieving the optimum biomass productivity for S. Obliquus ranged between 9 and 13. For an efficient nutrient removal, the proper N:P ratio is required to maintain a range between 9 and 13. (Bhatnagar et al., 2010) worked on chlorella minutissima and observed that it has the ability to grow so profusely in varied, contrasting situations and even can grew heterotrophically in the dark under acidic conditions. It shows that C. minutissima was a promising biomass builder in municipal water treatment. The approach of using microalgae for wastewater treatment is definitely a better option than supplying oxygen by pumping and other sources of energy.

5.5.4 Piggery/Swine Wastewater

Piggery wastewater composes of water, pig faces, pig urine and other substances. Due to the presence of high copper content, nitrogen, phosphorous, and other toxic material, it is necessary to pre-treat before redirecting it into an ecosystem. If such wastewater is directly discharged, it can pollute surface water as well as underground water. Different techniques and researches are going on to find the most effective way for the treatment of piggery wastewater. An Italian research team tried a full-scale sequencing batch reactor (SBR) system for the treatment, and a Swedish research team also used an instrumentation control and automation (ICA) for the wastewater treatment. Both techniques have shown effective results for cleaning the wastewater. However, using microalgae is also one of most effective and environmentally friendly ways to treat wastewater. (Markou and Georgakakis, 2011) studied that filamentous cyanobacterial (blue green algae) appeared to be suitable for the removal of organic and inorganic pollutants from wastewater because they produce biomass in satisfactory quantity and can be harvested easily. This process also includes a drawback that wastewater composition is not certain. It is seasonal. So the cultivation of algae varies seasonally. But this method can be considered as environmentally friendly and economical for the treatment.

Chlorella vulgaris microalgae shows the potential removal of nitrogen and phosphorous from piggery wastewater effluent. It was observed that the growth rate of plants increases with decreases in the concentration of piggery wastewater in the culture media independent of the diluents type. The microalgae and 20% concentrated wastewater diluted with distilled water have shown high potential removal by (Abou-Shanab et al., 2013; Ji et al., 2015) and conducted experiment on six different species of microalgae. The dry biomass of Ourococcusmultisporus, Nitzschia cf. pusilla, Chlamydomonasmexicana, Scenedesmus obliquus, Chlorella vulgaris, and Micractiniumreisseri were collected, and it was observed that C. Mexicana is the most favorable for the removal of pollutants from wastewater; thus, it can be considered as the best alternative for wastewater treatment.

5.5.5 Industrial Wastewater

In the past, many decades of rapid growth in industrialization occurred due to people's increased demand for many products to meet daily needs. Wastewater from the industries has an adverse effect on the ecosystem. There are various methods of treatment of this wastewater, and one of them is using microalgae. Microalgal treatment has been proved economical as well as effective because of its varied properties. Microalgal species of the genera chlorella, scenedesmus, and some cyanobacteria are the most employed species in various wastewater treatments due to their higher growth rate, environmental tolerance, and lipid and/or starch accumulation potential (Wang et al., 2016). The cocultivation of algae and bacteria can also stimulate algal growth and trigger carbohydrate and/or lipid accumulation, thereby enhancing the metal removal. Microalgae species with high metal sorption ability is crucial in achieving high metal elimination efficiency. They effectively eliminate heavy metals, nitrogen/ phosphates, and toxic components from the effluents. Because of the metabolic flexibility of microbial, i.e. their ability to perform photoautotrophic, mixotrophic, or heterotrophic metabolism, they are promising biological systems for treating toxic effluents. Microalgal biofuel production systems with wastewater treatment are effective as microalgae are known to effectively eliminate a variety of nutrients/pollutants in wastewater.

Algal pigments and photosynthesis operations have the ability to adapt to environmental conditions. It needs light for photochemical phase to produce ATP (adenosine triphosphate). Coenzymes are used as energy carriers in the cells of an organisms and dark for biochemical phase synthesis for molecular growth. Heavy metals are potent inhibitors of photosynthesis, so microalgae species, including C. sorokiniana and R. basilensis, are also effective in removing heavy metals from solution (Munoz et al., 2006). B. braunii, C.saccharophila, and P. carterae have been effective in treating effluents containing low concentrations of nitrogen and phosphorous. In addition, sargassum cymosum have been used for biosorption and bioaccumulation of heavy metals for industrial sludge. The microalgae can play a crucial role in producing the bioenergy if integrated with the remediation of wastewater while growing biomass for biofuel feedstock.

5.5.6 Pharmaceutical Wastewater

Pharmaceutical effluents are released into the environment by domestic, hospital, and pharmaceutical industries. This is serious hazard because of their direct effects on humans and ecotoxicity. Some of the commonly known drugs are used for experimental analysis; they include paracetamol used as an antipyretic drug; ibuprofen, an anti-inflammatory drug used for pain relief, fever, and inflammation; olanzapine, an antipsychotic drug used for schizophrenia. Due to low concentration of drugs in wastewater and their properties, such as hydrophilicity, solubility, volatility, and biodegradability, the common conventional treatment methods are not effective for their removal. So, green technologies are tried, and one is the microalgal degradation. Microalgae can remove personal-care compounds, such as triclosan from wastewater; it depends on microalgae strain and operating conditions. The removal process of personal-care compounds could be accumulation, adsorption, and degradation either extracellular or intracellular. Some microalgae has the ability to uptake and separate organic substrates from wastewater due to seasonal variation like diclofenac, benzothiazola, and OH-Benzothiazole etc. Some other personal-care compounds, also due to temperature variation, can be removed from wastewater; they include ibuprofen, acetaminophen, caffeine etc. Nannochloropsis sp. microalgae is very effective in removing paracetamol and ibuprofen due to its high growth rate, resilience, adaptation to a wide range of growth media, halo tolerance, and accumulation of large amount of lipids. (Xiong et al., 2016) studied microbial consortium containing Chlorella sp., and Scenedesmus sp. can successfully remove 20% of CBZ (carbamazepine most commonly used in pharmaceutical industries).

Wirth et al. (2020) reported the removal capability of pH, BOD, phosphate, ammonia, carbon, and nitrogen using chlorella vulgaris microalgae and its phycosphere, as shown in Figure 5.3. The experimental studies were done for four days with fermentation effluent (FE), municipal wastewater (MW), and chicken manure supernatant (CMS) wastewater. Removal of phosphate and total nitrogen depends on light intensity. The phosphate-removal rate was observed as higher in CMS (0.20 $mMday^{-1}$) but in fermentation effluent (FE), municipal wastewater (MW) was observe only (0.02 mM day^{-1}). Nitrogen removal rates were observed 0.78 mM day^{-1} in FE, 0.32 $mMday^{-1}$ in MW, and 2.46 $mMday^{-1}$ in four days, and the same removal trains were observed in ammonia content (2.44 $mMday^{-1}$ in CMS, 0.77 $mMday^{-1}$ in FE, and 0.31 $mMday^{-1}$ in MW), as shown in Figure 5.3d. During the experiments, water pH is increased as expected, as shown in Figure 5.3a, and BOD variation, as shown in Figure 5.3b. The biomass production is highest in CPM as compared to other wastewater and least in municipal wastewater, as shown in Figure 5.3g.

5.6 CHALLENGES IN INTEGRATING WASTEWATER TREATMENT AND ALGAE BIOMASS PRODUCTION

The main challenges of cultivation of microalgae will mostly depend on selection of appropriate wastewater, pretreatment methods for wastewater, screening, selection and weather condition for microalgae to sustain and grow in such environment.

1. **Selection of wastewater:** The main nutrient requirements for algal growth in water are nitrogen, phosphorus, and potassium. In the production of biomass in the presence of photosynthesis, algae takes nutrients along with CO_2. The N:P ratio in water plays an important role in growing algae, and it should be in an optimum range (Dodds and Smith, 2016). The harvesting cost of algae biomass accounts for a total of 30% of the production cost. The N:P ratio for some different wastewater like Dairy-14, Paper mill-41, Textile-11, Winery-5, Poultry feedlot-36, Beef cattle feedlot-10, and

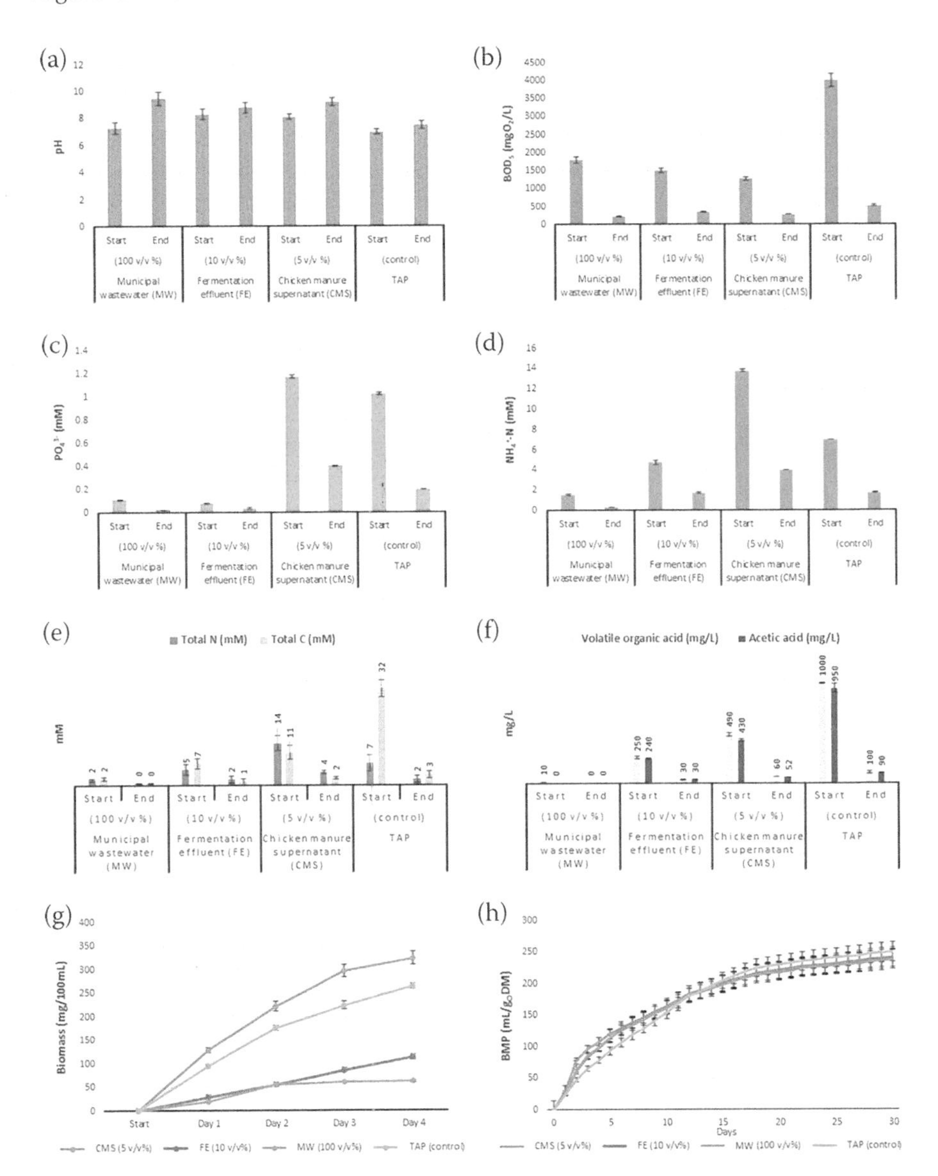

FIGURE 5.3 Summary of microalge-bacterial bioremediation and cultivation efficiency on different types of wastewater. (a) Results of pH measurement. (b) Results of biological oxygen demand calculation. (c) Total Phosphate measurements. (d) Ammonium ion measurement data. (e) Total carbon and nitrogen contents. (f) Volatile organic acid (VOAs) and acetic acid concentrations. (g) Biomass growth dynamics over time (days) (h) Cumulative biological methane potential of cultivated biomass. (adopted from Wirth et al., 2020).

Domestic waste-11 to 13 (Christenson and Sims, 2011.) The optimum value of N/P in municipal wastewater treatment using microalgae is 5 to 30 for biomass as well as nutrient removal (1). The growth of Chlorella pyrenoidosa in soybean processing (water) observed faster removal of N over P, Scenedesmus sp. Required an N:P ratio of growth deprived of limitation by either nutrient. The range of 12 to 18 algae growing in an environment faced incessant inadequacy. Hence, the rate of N elimination is always higher than P.

2. **Wastewater pretreatment method:** Wastewater requires pretreatment methods such as filtration, autoclaving, UV-irradiation, and dilution before growing microalgae in water (Cho et al., 2011). This entire process is challenging, and this process will take more time and space. Out of several methods, filtration methods are the most effective for large applications. Regular maintenance is required to clean tanks for the cultivation of algae to prevent the contamination by the other microorganisms.
3. **Screening and selection of microalgae:** The selection of suitable microalgae is one of the main challenges working for biomass production. The screened and isolated microalgae are used to successfully integrate microalgae cultivation with advanced treatment of wastewater because those isolates are already well adapted to the condition in wastewater.
4. **Weather condition:** Microalgae biomass for large-scale production in open systems reaches up to 0. 5 g/L, which could be mostly attributed to light dispersion at the top layer and for deeper culture would reduce biomass density. To increase the large-scale biomass production of microalgae would optimize water flow velocity. This results in improved mixing and light utilization within the tank.
5. The biomass production could improve by properly maintaining the concentration of phosphorous and nitrogen in wastewater. If the pH level in wastewater is low or high, it should be maintained within the range for efficient microalgae bioremediation. High concentration of ammonia in wastewater could be toxic for many microalgae strains. So, we need to maintain regular intervals within the permissible range. For cell growth of microalgae required sunlight, this will depend on weather conditions. Low growth rate of microalgae occurs in the winter season due to low intensity of light, so, weather conditions also depend on biomass production.

5.7 CONCLUSION

Microalgae is one of the green technologies developed for wastewater treatment, as well as different potential application of biomass. This chapter discussed a detailed description of algal biomass production and treatment of nutrients-rich wastewater, such as municipal wastewater, industrial wastewater, pharmaceutical wastewater, and textile wastewater. The biomass obtained from different wastewater treatments can be used in different applications, such as energy production, animal feed production, production of chemicals, biofertilizer, and human food. The large production of microalgae in open ponds is currently available, and work has been done to some extent in cold ponds, but more focus is needed on closed ponds for large-scale production.

ACKNOWLEDGEMENT

The authors acknowledge the Collaborative Research Scheme (CRS) under Technical Education Quality Improvement Plan (TEQIP-III) for funding grant of project ID: 1–5764058221 for research work.

REFERENCES

Abou-Shanab, JiMK, Kim, H.C., Paeng, K.J., and Jeon, B.H.. (2013). Microalgal species growing on piggery wastewater as a valuable candidate for nutrient removal and biodiesel production. *J Environ. Manag.*, *115*, 257–264. doi:10.1016/j.jenvman.2012.11.022.

Acien, F.G., Fernández-Sevilla, J.M., and Molina-Grima, E. (2017). Microalgae: The basis of making sustainability. In *Case study of innovative projects – Successful real cases*. doi:10.5772/67930.

Al Hamadi, A., Uraz, G., Katırcıoğlu, H., and Osmanağaoğlu, Ö. (2017). Adsorption of Azo dyes from textile wastewater by Spirulina Platensis. *Eurasian Journal of Environmental Research*, *1*, 19–27. https://dergipark.org.tr/en/pub/ejere/issue/31884/334653

AlFadhly, Nawal K. Z., Alhelfi, Nawfal, Altemimi, Ammar B., Verma, Deepak Kumar, Cacciola, Francesco, & Narayanankutty, Arunaksharan (2022). Trends and Technological Advancements in the Possible Food Applications of Spirulina and Their Health Benefits: A Review. Molecules, 27, 5584. doi:10.3390/molecules27175584.

An, J-Y., Sim, S-J, Lee, J.S., and Kim, B.W. (2003). Hydrocarbon production from secondarily treated piggery wastewater by the green alga Botryococcus braunii. *J Appl Phycol.*, *15*, 185–191. doi:10.1023/A:1023855710410.

Andreotti, V., Solimeno, A., Rossi, S., Ficara, E., Marazzi, F., Mezzanotte, V., García, J. (2020). Bioremediation of aquaculture wastewater with the microalgae Tetraselmis suecica: Semi-continuous experiments, simulation and photo-respirometric tests. *Sci. Total Environ*, *738*, 139859. doi:10.1016/j.scitotenv.2020.139859.

Arteaga, L.C., Zavaleta, M.P., Eustaquio, W.M., and Bobadilla, J.M. (2018). Removal of aniline blue dye using live microalgae *Chlorella vulgaris*. *Journal of Energy & Environmental Sciences*, *2*, 6–12. doi:10.32829/eesj.v2i1.51.

Arbib Z., Ruiz, J., Álvarez-Díaz, P., Garrido-Pérez, C., Barragan, J., and Perales, J.A. (2013). Long term outdoor operation of a tubular airlift pilot photobioreactor and a high rate algal pond as tertiary treatment of urban wastewater. *Ecol Eng.*, *52*, 143–153 doi:10.1016/j.ecoleng.2012.12.089.

Becker, E.W. (1994). *Microalgae: Biotechnology and microbiology*. Cambridge University Press, Cambridge. doi:10.1017/s0014479700025126.

Behl, K., SeshaCharan, P., Joshi, M., Sharma, M., Mathur, A., Kareya, M.S., Jutur, P.P., Bhatnagar, A., and Nigam, S. (2020). Multifaceted applications of isolated microalgae Chlamydomonas sp. TRC-1 in wastewater remediation, lipid production and bioelectricity generation. *Bioresource Technology*, *304*, 122993. doi:10.1016/j.biortech.2020.122993.

Bhatia, D., Sharma, N.R., Singh, J., and Kanwar, R.S. (2017). Biological methods for textile dye removal from wastewater: A review. *Critical Reviews in Environmental Science and Technology*, *47*(19), 1836–1876. doi:10.1080/10643389.2017.1393263.

Bhatnagar A., Bhatnagar, M., Chinnasamy, S., and Das, K. (2010). Chlorella minutissima-apromising fuel alga for cultivation in municipal wastewaters. *Appl. Biochem. Biotech.*, *161*, 523–536. doi:10.1007/s12010-009-8771-0.

Cañizares-Villanueva, R.O., Domínguez, A.R., Cruz, M.S., and Ríos-Leal, E. (1995). Chemical composition of cyanobacteria grown in diluted, aerated swine wastewater. *Bioresource Technology*, *51*, 111–116. doi:10.1016/0960-8524(94)00099-M.

Cecal, A., Humelnicu, D., Rudic, V., Cepoi, L., Ganju, D., and Cojocari, A. (2012). Uptake of uranyl ions from uranium ores and sludges by means of Spirulina platensis, Porphyridium cruentum and Nostok linckia alga. *Bioresource Technology, 118*, 19–23. doi:10.1016/j.biortech.2012.05.053.

Cechinel, M.A.P., Mayer, D.A., Pozdniakova, T.A., Mazur, L.P., Boaventura, R.A.R., de Souza, A.A.U., de Souza, S.M.A.G.U., and Vilar, V.J.P. (2016). Removal of metal ions from a petrochemical wastewater using brown macro-algae as natural cationexchangers. *Chem. Eng. J., 286*, 1–15. doi:10.1016/j.cej.2015.10.042.

Chai, Wai Siong, Tan, Wee Gee, Halimatul Munawaroh, Heli Siti, Gupta, Vijai Kumar, Ho, Shih-Hsin, and Show, Pau Loke (2021). Multifaceted roles of microalgae in the application of wastewater biotreatment: A review. *Environmental Pollution*, 269, 116236. doi:10.1016/j.envpol.2020.116236.

Chan, A., Salsali, H., and Mc Bean, E. (2013). Heavy Metal Removal (Copper and Zinc) in Secondary Effluent fromWastewater Treatment Plants by Microalgae. *ACS Sustain. Chem. Eng., 2*, 130–137. doi:10.1021/sc400289z.

Chakraborty, S., and Santra, S.C. (2008). Biochemical composition of eight benthic algae collected from Sunderban. *Indian Journal of Marine Sciences, 37*, 329–332. http://nopr.niscair.res.in/handle/123456789/2057

Charumathi, D., and Das, N. (2012). Packed bed column studies for the removal of synthetic dyes from textile wastewater using immobilised dead C. tropicalis. *Desalination, 285*, 22–30, doi:10.1016/j.desal.2011.09.023.

Chisti Y. (2007). Biodiesel from microalgae. *Biotechnol Adv., 25*, 294–306. doi:10.1016/j.biotechadv.2007.02.001.

Christenson L., and Sims R. (2011). Production and harvesting of microalgae for wastewater treatment, biofuels, and bioproducts. *Biotechnol. Adv., 29*, 686–702. doi:10.1016/j.biotechadv.2011.05.015.

Cho, S., Luong, T.T., Lee, D., Oh, Y.K., and Lee, T. (2011). Reuse of effluent water from a municipal wastewater treatment plant in microalgae cultivation for biofuel production. *Bioresource Technology, 102*, 8639–8645. doi:10.1016/j.biortech.2011.03.037.

Chong, A.M.Y., Wong, Y.S., and Tam, N.F.Y. (2000). Performance of different microalgal species in removing nickel and zinc from industrial wastewater. *Chemosphere, 41*, 251–257. doi:10.1016/S0045-6535(99)00418-X.

Daneshvar, N., Ayazloo, M., Khataee, A.R., and Pourhassan, M. (2007). Biological decolorization of dye solution containing Malachite Green by microalgae Cosmarium sp. *Bioresource Technology, 98*, 1176–1182. doi:10.1016/j.biortech.2006.05.025.

Dodds, W.K., and Smith, V.H. (2016). Nitrogen, phosphorus, and eutrophication in streams. *Inland Waters, 6*, 155–164. doi:10.5268/IW-6.2.909.

De Godos, I., Arbid, Z., Lara, E., Cano, R., Muñoz, R., and Rogalla, F. (2017). Wastewater treatment in algal systems. In Lema, J.M., and Suarez, S. (Eds.),*Innovative wastewater treatment and resource recovery technologies. Impacts on energy, economy and environment* (pp. 76–95). IWA Publishing, London. doi:10.2166/9781780407876_0076.

Doğar, C., Gürses, A., Açıkyıldız, M. and Özkan, E. (2010). Thermodynamics and kinetic studies of biosorption of a basic dye from aqueous solution using green algae *Ulothrix* sp. *Colloids and Surfaces B: Biointerfaces, 76*, 279–285. doi:10.1016/j.colsurfb.2009.11.004.

Do, J.M., Jo, S.W., Kim, I.S., Na, H., Lee, J.H., Kim, H.S., and Yoon, H.S. (2019). A feasibility study of wastewater treatment using domestic microalgae and analysis of biomass for potential applications. *Water, 11*, 2294. doi:10.3390/w11112294.

Dotto, G.L., Lima, E.C., and Pinto, L.A.A. (2012). Biosorption of food dyes onto Spirulina platensis nanoparticles: Equilibrium isotherm and thermodynamic analysis. *Bioresource Technology*, 103, 123–130. doi:10.1016/j.biortech.2011.10.038.

Fernández, F.G.A., Sevilla, J.M.F., and Grima, E.M. (2013). Photobioreactors for the production of microalgae. *Rev. Environ. Sci. Biotechnol.*, *12*, 131–151. doi:10.1007/s11157-012-9307-6.

Gelebo, G., Tessema, L., Kehshin, K., Gebremariam, H., Gebremikal, E., Motuma, M., Ayele, A., Getachew, D., Benor, S., and Suresh, A. (2020). Phycoremediation of synthetic dyes in an aqueous solution using an indigenous Oscillatoria sp., from Ethiopia. *Ethiopian Journal of Sciences and Sustainable Development*, *7*, 14–20. doi:10.20372/ejssdastu:v7.i2.2020.186.

González, A.G., Oleg, S., Pokrovsky, J., Santana-Casiano, M., and González-Dávila, M. (2017). Bioadsorption of heavy metals. In Tripathi, B. , and Kumar, D. (Eds.), *Prospects and Challenges in Algal Biotechnology* (pp. 233–255). Springer, Singapore. doi:10.1007/978-981-10-1950-0_8.

Han, Y., Wen, Q., Chen, Z., and Li, P. (2011). Review of methods used for microalgal lipid-content analysis. *Energy procedia*, *12*, 944–950. doi:10.1016/j.egypro.2011.10.124.

Hu B., Min, M., Zhou, W.G., Li, Y.C., Mohr, M., Cheng, Y.L., Lei, H.W., Liu, Y.H., Lin, X.Y., Chen, P., and Ruan, R. (2012). Influence of Exogenous CO_2 on Biomass and Lipid Accumulation of Microalgae Auxenochlorella protothecoides Cultivated in Concentrated Municipal Wastewater. *Appl. Biochem. Biotech. 2012*, *166*, 1661–1673. doi:10.1007/s12010-012-9566-2.

Ishchi, T., and Sibi, G. (2020). Azo dye degradation by Chlorella vulgaris: optimization and kinetics. *International Journal of Biological Chemistry*, *14*, 1–7. doi:10.3923/ijbc.2020.1.7.

Ji, F., Liu, Y., Hao, R., Li, G., Zhou, Y., and Dong, R. (2014). Biomass production and nutrients removal by a new microalgae strain Desmodesmus sp. in anaerobic digestion wastewater. *Bioresource Technology*, *161*, 200–207. doi:10.1016/j.biortech.2014.03.034.

Ji, M.-K., Yun, H.-S., Park, S., Lee, H., Park, Y.-T., Bae, S., Ham, J., and Choi, J. (2015). Effect of food wastewater on biomass production by a green microalga Scenedesmus obliquus for bioenergy generation. *Bioresource Technology*, *179*, 624–628. doi:10.1016/j.biortech.2014.12.053.

Katiyar, R., Banerjee, S., and Arora, A. (2021). Recent advances in the integrated biorefinery concept for the valorization of algal biomass through sustainable routes. Biofuels, Bioproducts and Biorefining, 15, 879–898. doi:10.1002/bbb.2187.

Khalaf, M.A. (2008). Biosorption of reactive dye from textile wastewater by non-viable biomass of *Aspergillus niger* and *Spirogyra* sp. *Bioresource Technology*, *99*, 6631–6634. doi:10.1016/j.biortech.2007.12.010.

Khataee, A.R., Dehghan, G., Ebadi, E., and Pourhassan, M. (2010). Central composite design optimization of biological dye removal in the presence of macroalgae *Chara* sp, *CLEAN–Soil, Air, Water*, *38*, 750–757. doi:10.1002/clen.200900295.

Kong, Q.X., Li, L., Martinez, B., Chen, P., and Ruan, R. (2010). Culture of microalgae Chlamydomonas reinhardtii in wastewater for biomass feedstock production. *Appl. Biochem. Biotechnology*, *160*, 9–18. doi:10.1007/s12010-009-8670-4.

Kumar, R., and Goyal, D. (2010). Waste water treatment and metal (Pb^{2+}, Zn^{2+}) removal by microalgal based stabilization pond system. *Indian J. Microbiol*, *50*, 34–40. doi:10.1007/s12088-010-0063-4.

Li, H., Watson, J., Zhang, Y., Lu, H., and Liu, Z. (2020). Environment-enhancing process for algal wastewater treatment, heavy metal control and hydrothermal biofuel production: A critical review. *Bioresource Technology*, *298*, 122421. doi:10.1016/j.biortech.2019.122421.

Li, Y., Chen, Y.F., Chen, P., Min, M., Zhou, W., Martinez, B., Zhu, J., and Ruan, R. (2011a). Characterization of a microalga Chlorella sp. well adapted to highly concentrated municipal wastewater for nutrient removal and biodiesel production. *Bioresource Technology*, *102*, 5138–5144. doi:10.1016/j.biortech.2011.01.091.

Li, Y., Yang, X., and Geng, B. (2018). Preparation of immobilized sulfate-reducing bacteria-microalgae beads for effective bioremediation of copper-containing wastewater. *Water. Air. Soil Pollut.*, *229*, 54. doi:10.1007/s11270-018-3709-1.

Markou, G., and Georgakakis, D. (2011). Cultivation of filamentous cyanobacteria (blue-green algae) in agro-industrial wastes and wastewaters. *Applied Energy*, *88*, 3389–3401. doi:10.1016/j.apenergy.2010.12.042.

Mahini, R., Esmaeili, H., and Foroutan, R. (2018). Adsorption of methyl violet from aqueous solution using brown algae *Padina sanctae-crucis*. *Turkish Journal of Biochemistry*, *43*, 623–631. doi:10.1515/tjb-2017-0333.

Mehta, S.K., and Gaur, J.P. (2005). Use of algae for removing heavy metal ions from wastewater: Progress and prospects. *Crit. Rev. Biotechnol.*, *25*(3), 113–152. doi:10.1080/07388550500248571.

Molinuevo-Salces, B., Mahdy, A., Ballesteros, M., and González-Fernández, C. (2016). From piggery wastewater nutrients to biogas: microalgae biomass revalorization through anaerobic digestion. *Renew Energy*, *96*, 1103–1110. doi:10.1016/j.renene.2016.01.090.

Moheimani, N.R. (2013). Inorganic carbon and pH effect on growth and lipid productivity of Tetraselmis suecica and Chlorella sp (Chlorophyta) grown outdoors in bag photobioreactors. *J. Appl. Phycol.*, *25*, 387–398. doi:10.1007/s10811-012-9873-6.

Munoz, R., Alvarez, M.T., Munoz, A., Terrazas, E., Guieysse, B., Mattiasson, B. (2006). Sequential removal of heavy metals ions and organic pollutants using an algalbacterial consortium. *Chemosphere*, *63*(6), 903–911. doi:10.1016/j.chemosphere.2005.09.062.

Naji, N.S., and Salman, J.M. (2019). Effect of temperature variation on the efficacy of *Chlorella vulgaris* in decolorization of Congo red from aqueous solutions. *Biochemical and Cellular Archives*, *19*, 4169–4174. doi:10.35124/bca.2019.19.2.4169.

Parsy, A., Sambusiti, C., Baldoni-Andrey, P., Elan, T., and Périé, F. (2020). Cultivation of Nannochloropsis oculata in saline oil & gas wastewater supplemented with anaerobic digestion effluent as nutrient source. *Algal Res.*, *50*, 101966. doi:10.1016/j.algal.2020.101966.

Pereira, S., Micheletti, E., Zille, A., Santos, A., Moradas-Ferreira, P., Tamagnini, P., and De Philippis, R. (2011). Using extracellular polymeric substances (EPS)-producing cyanobacteria for the bioremediation of heavy metals: Do cations compete for the EPS functional groups and also accumulate inside the cell *Microbiology*, 157, 451–458. doi:10.1099/mic.0.041038-0.

Phang, S.M., Miah, M.S., Yeoh, B.G., and Hashim, M.A. (2000). Spirulina cultivation in digested sago starch factory wastewater. *J. Appl. Phycol.*, *12*, 395–400. doi:10.1023/A:1008157731731.

Pluz, O., and Gross, W. (2004). Valuable products from biotechnology of microalgae. *Appl. Microbiol. Biotechnol.*, *65*, 635–648. doi:10.1007/s00253-004-1647-x.

Pradhan, D., Sukla, L.B., Mishra, B.B., and Devi, N. (2019). Biosorption for removal of hexavalent chromium using microalgae Scenedesmus sp. *J. Clean. Prod.*, *209*, 617–629. doi:10.1016/j.jclepro.2018.10.288.

Pratiwi, D., Prasetyo, D.J., and Poeloengasih, C.D. (2019). Adsorption of methylene blue dye using marine algae *Ulva lactuca*. *IOP Conference Series: Earth and Environmental Science*, *251*, 012012, 2019. doi:10.1088/1755-1315/251/1/012012.

Priyadarshani, I., and Rath, B. (2012). Commercial and industrial applications of micro algae – A review. *J. Algal Biomass Utln*, *3*, 89–100.

Rajeshkannan, R., Rajasimman, M., and Rajamohan, N. (2010). Optimization, equilibrium and kinetics studies on sorption of Acid Blue 9 using brown marine algae Turbinaria conoides. *Biodegradation*, 21, 713–727. doi:10.1007/s10532-010-9337-0.

Rani, S., Gunjyal, N., Ojha, C.S.P., and Singh, R.P. (2021). Review of challenges for algae-based wastewater treatment: Strain selection, wastewater characteristics, abiotic, and biotic factors. *Journal of Hazardous, Toxic, and Radioactive Waste, 25*(2), 03120004: doi:10.1061/(ASCE)HZ.2153-5515.0000578.

Ricketts, T.R. (1966). On the chemical composition of some unicellular algae. *Phytochemistry, 5*, 67–76. doi:10.1016/S0031-9422(00)85082-7.

Salama E.I., Kurade, M.B., Abou-Shanab, R.A.I., El-Dalatony, M.M., Yanga, Il-S., Minc, B., and Jeona, B.H. (2017). Recent progress in microalgal biomass production coupled with wastewater treatment for biofuel generation. *Renewable and Sustainable Energy Reviews, 79*, 1189–1211. doi:10.1016/j.rser.2017.05.091.

Salas, L.M.L., Castrillo, M., and Martinez, D. (2013). Effects of dilution rates and water reuse on biomass and lipid production of Scenedesmus obliquus in a two-stage novel photobioreactor. *Bioresource Technology, 143*, 344–352. doi:10.1016/j.biortech.2013.06.007.

Sarkar, P., and Dey, A. (2021). Phycoremediation – An emerging technique for dye abatement: An overview. *Process Safety and Environmental Protection, 147*, 214–225. doi:10.1016/j.psep.2020.09.031.

Shanab, S., Essa, A., and Shalaby, E. (2012). Bioremoval capacity of three heavy metals by some microalgae species (Egyptian isolates). *Plant Signaling and Behaviour, 7*, 392–399. doi:10.4161/psb.19173.

Sheehan, J., Dunahay, T., Benemann, J., and Roessler, P. (1998). A look back at the U.S. Department of Energy's Aquatic Species Program. *Program*, 328, 1–249. doi:10.2172/15003040.

Sibi, G. (2016). Biosorption of chromium from electroplating and galvanizing industrial effluents under extreme conditions using Chlorella vulgaris. *Green Energy Environ, 1*, 172–177. doi:10.1016/j.gee.2016.08.002.

Solovchenko, A., Pogosyan, S., Chivkunova, O., Selyakh, I., Semenova, I., Voronova, E., Scherbakov, P., Konyukhov, I., Chekanov, K., Kirpichnikov, M., and Lobakova, E. (2014) Phycoremediation of alcohol distillery wastewater waitha novel Chlorella sorokiniana strain cultivated in a photobioreactor monitored on-line via chlorophyll fluorescence. *Algal Res, 6*, 234–241. doi:10.1016/j.algal.2014.01.002.

Sydney, E.B., Sturm, W., de Carvalho, J.C., Thomaz-Soccol, V., Larroche, C., Pandey, A., and Soccol, C.R.. (2010). Potential carbon dioxide fixation by industrially important microalgae. *Bioresour Technology, 101*, 5892–5896. doi:10.1016/j.biortech.2010.02.088.

Travieso, L., Cañizares, R.O., Borja, R., Benítez, F., Domínguez, A.R., Dupeyrón, R., and Valiente, Y.V. (1999). Heavy metal removal by microalgae. *Bull. Environ. Contam. Toxicol., 62*, 144–151. doi:10.1007/s001289900853.

Tuantet, K., Temmink, H., Zeeman, G., Janssen, M., Wijffels, R.H., and Buisman, C.J. (2014). Nutrient removal and microalgal biomass production on urine in a short light-path photobioreactor. *Water Res., 55*, 162–174. doi:10.1016/j.watres.2014.02.027.

Uebel, L.S., Costa, J.A.V., Olson, A.C., and Morais, M.G. (2019). Industrial plant for production of *Spirulina* sp. *LEB 18, Braz. J. Chem. Eng. 35*. doi:10.1590/0104-6632.20180361s20170284

Wang B., Lan, C.Q., and Horsman, M. (2012). Closed photobioreactors for production of microalgal biomasses. *Biotechnology Advances, 30*, 904–912. doi:10.1016/j.biotechadv.2012.01.019.

Wang, Y., Ho, S.H., Cheng, C.L., Guo, W.Q., Nagarajan, D., Ren, N.Q., Lee, D.J., and Chang, J.S. (2016). Perspectives on the feasibility of using microalgae for industrial wastewater treatment. *Bioresource Technology, 222*, 485–497. doi:10.1016/j.biortech.2016.09.106.

Wirth R., Bernadett, , Tamás, B., Prateek, S., Gergely, L., Zoltán, B., K. Kornél L., and Gergely, M. (2020). Chlorella vulgaris and its Phycosphere in Wastewater: Microalgae-bacteria interactions during nutrient removal. *Frontiers in Bioengineering and Biotechnology, 8*, 1108–1123.doi:10.3389/fbioe.2020.557572.

Wong, J.P.K., Wong, Y.S., and Tam, N.F.Y. (2000). Nickel biosorption by two chlorella species, C. Vulgaris (a commercial species) and C.Miniata (a local isolate). *Bioresour. Technol.*, *73*, 133–137. doi:10.1016/S0960-8524(99)00175-3.

Wu, L.F., Chen, P.C., Huang, A.P., and Lee, C.M. (2012). The feasibility of biodiesel production by microalgae using industrial wastewater. *Bioresour Technology*, *2012*(113), 14–18. doi:10.1016/j.biortech.2011.12.128.

Xiong, J.Q., Kurade, M.B., Abou-Shanab, R.A.I., Ji, M.K., Choi, J., Kim, J.O., and Jeon, B.H. (2016). Biodegradation of carbamazepine using freshwater microalgae Chlamydomonas mexicana and Scenedesmus obliquus and the determination of its metabolic fate. *Bioresource Technology*, *205*, 183–190. doi:10.1016/j.biortech.2016.01.038.

Zhou W., Li, Y., Min, M., Hu, B., Chen, P., and Ruan, R. (2011). Local bioprospecting for high-lipid producing microalgal strains to be grown on concentrated municipal wastewater for biofuel production. *Bioresour Technology*, *2011*(102), 6909–6919. doi:10.1016/j.biortech.2011.04.038.

6 Photobioreactor in Wastewater Treatment: Design and Operational Features

Ranjana Das and Chiranjib Bhattacharjee
Chemical Engineering Department, Jadavpur University, Kolkata, West Bengal, India

CONTENTS

6.1 INTRODUCTION

The photobioreactor model originated from the concept of controlling the issue of global warming caused by CO_2 emissions, which has become the key impediment to sustainable social and economic development. One of the most promising technologies involving CO_2 fixation is to grow microalgae in photobioreactors (PBR), through which its aeration with CO_2-is enriched with gas from coal-fired flue gas. PBR is an essential device for microalgae cultivation as it offers suitable conditions, such as appropriate light, carbon source, nutrients, pH, and temperature (Sun et al., 2016). The PBR is a multifaceted device not only for CO_2 fixation but also emerging as a potential route of water treatment. Photobioreactors (PBRs) have a major role in cultivating phototrophic microorganisms to fix CO_2 and produce targeted products. A PBR is a bioreactor that utilizes a light source to cultivate phototrophic microorganisms. These cultivated organisms by the photosynthesis route generate biomass from light and carbon dioxide and include macroalgae, microalgae, cyanobacteria, and purple bacteria plants and mosses. In the artificial environment of a photobioreactor, specific conditions are carefully controlled for respective species, as per selection, allowing much higher growth rates and purity levels than natural habitats. Both the possibilities exist for generation of the phototropic biomass from carbon dioxide sources and nutrient-rich wastewater, but in the present article, the prospect of PBR in terms of water treatment has been

DOI: 10.1201/9781003165101-6

discussed in detail. PBR application in wastewater recuperation, the design parameters to control being extremely relevant, numerous researches have already been conducted to establish the feasibility issue of PBR (Huang et al., 2017; Hao et al., 2018; Sun et al., 2018; Fu et al., 2019; Nhat et al., 2019). Despite enormous researches, a feasible PBR design that can be used commonly for large-scale culturing of microalgae has not yet been achieved as the choice of PBR depends on the microalga species, desired products, yield value, and, most importantly, on maximizing the light penetration into the reactor (Nhat et al., 2019) with a high surface-to-volume ratio and the use of transparent building materials to maximise light penetration. In the case of lower surface-to-volume ratio design of PBR, internal LED-lighting arrangements are maintained to improve light penetration and distribution. Designing PBRs is still challenging, and most of the reactors have been designed and scaled up using semi-empirical approaches. No appropriate types of PBRs are available for mass cultivation due to the reactors' high capital and operating costs and short lifespan, which are mainly due to a current lack of appropriate understanding of the coupled behavior of light, hydrodynamics, mass transfer, and cell growth in efficient reactor design. Photobioreactors is a cheap and self-sustaining process in which wastewater is treated with active aerobic microorganisms, resulting in detoxification of effluent and production of carbon dioxide that promotes algal growth with the potential of producing algal fuel. This specialised bioreactor, with a given set of conditions, can grow any species of algae, cyanobacteria, seaweed, or plant cell efficiently. Applications for photobioreactors are not only limited to carbon capture and wastewater treatment but also with diverse applications involving i) culturing macro and micro algae, ii) cyanobacteria and purple bacteria, iii) seaweed, plant cells, and bryophytes, iv) wastewater remediation, v) hydrogen gas production (biorefineries), vi) algal alcohols and oils as biofuels, vii) production of biomolecules (polysaccharides, amino acids, isoprenoids, phenols), viii) bioplastics, organic renewable replacements for petrochemical products, ix) aquaculture (animal feed, human nutrition, value added products), x) Carbon capture (sinking carbon to mitigate climate change), xi) pharmaceutical applications (drug discovery, bioactive molecules, vaccine production). For the industrial production of biochemical components involving sustainable human activities, biofuel production, researches of bioreactor design have become explicitly relevant. In practice, industrial microalgae productions for biofuel production requires large-scale culturing in open systems like open pond and raceway ponds being an economic approach in terms of construction, operation, and maintenance; but they may be obstructed by several issues like contamination, evaporation, variation of environmental condition, and extensive land requirements. Closed PBR systems can eliminate contamination and evaporation issues and achieve higher biomass concentrations at the expense of high capital cost, technical hitches of scaling up, and high shear stress. However, because of high yield value, easy control of the growth condition researches are involved in designing efficient, low-cost PBR systems. Waste effluents are the main entrance pathways for a broad variety of organic micropollutants into the aquatic environment, even after tertiary and/or advanced treatments such as UV radiation, membrane bioreactors (MBR), reverse osmosis (RO), or nanofiltration (NF) process, the needed final treatment for being ecosafe effluent (Biel-Maeso et al., 2018; Mamo et al., 2018; Racar et al., 2020; García-Galán et al., 2021). Nature-inspired microalgae-based treatments are economically feasible and a potential alternative to conventional WWTPs for the human race. The process principle of PBR systems has the dual ability of treating wastewater and producing biomass, which can further be processed to produce profitable bioenergy like biogas (Zhu, 2015), value-added products like biofertilizers, pigments, and bioplastics (Rueda et al., 2020). In spite of several advantageous features and the availability of PBR designs, utilization of PBR on an industrial scale is mainly constrained by the economics and complexity of operation. This article involves details of the photobioreactor application for wastewater treatment, emphasizing the salient design aspect of the device and operational features. It points out the relevance of various parameters in designing PBR, various configurations of PBR, and design aspects of novel type PBR for specific application.

6.2 COMMERCIAL PHOTOBIOREACTORS APPLICATIONS IN WASTEWATER TREATMENT

The origin of the photobioreactor concept has been started till development of the open algal pond (Palmer, 1974). These open raceway ponds use algal consortia along with nutrients to circulate around the raceway track to provide necessary mixing as well as transportation of culture from input to output. Up-scaling of open pond systems requires land, and such systems would be best suitable for wastewater treatment, particularly in village areas with low land costs, least monitoring, low maintenance, and low power consumption. Photobioreactors with compact design and sophisticated process control have emerged as potential alternatives to the open raceway pond systems for wastewater treatment. Some commercial PBR configurations in use have been mentioned in this chapter practiced for wastewater treatment. (a) Tank or box configurations, (b) cylindrical or columnar configuration, (c) tubular configurations, (d) helical-shaped photobioreactor, (e) conical tubular photobioreactor, (f) horizontal tubular photobioreactor, (g) vertical-coiled tubular photobioreactor, (h) flat-plate or flat-panel reactors (Ashok et al., 2019) are some of the industrially tested PBR assemblies already reported.

The development of the photobioreactors (PBRs) is recently noticeable as cutting-edge technology, while the correlation of PBs' engineered elements, such as modellings, configurations, biomass yields, operating conditions, and pollutants removal efficiency still remains complex and unclear necessitating a systematic understanding of PBRs. Photobioreactor configuration has been reported as a critical parameter in generating the eukaryotes community, while the bacterial community was reported to be influenced strongly by both configuration and medium nature. Choice of the wavelength of light in PBR system has also been reported as a critical factor for achieving optimal algal-bacterial consortium growth. The 'tank and box' configuration of the photoreactor is a typical closed stirred tank reactor. Closed tank systems are generally made of glass. The reactor is made of plexiglas, or polyethylene where mixing can be done either by stirring, mixing paddles, or air bubbling using wastewater as feed (Huang et al., 2017). In closed tank reactors, air sparging, degasification, mixing, and illumination costs are some of the factors of concern, as well as difficulties in uniform distribution of light and regular cleaning and maintenance of the reactor. A cylindrical airlift and cylindrical air bubbling reactors have been in use with their salient features. Higher biomass growth rates have been reported in columnar reactors, with fewer contamination problems. However, contamination in such reactors is higher than closed tubular reactors and lesser than open tank systems. These reactors require greater care for mixing provisions by allowing central particles to come closer to the periphery to receive light; partial aerobic and anaerobic portions may get created due to trapped gases, unavailability of proper lighting, and mixing conditions.

The major advantage of such systems is lesser land. Requirements are due to its vertical extent and uniform distribution of light and mixing due to its circular shape. In addition, by virtue of its simple design, it allows ease in maintenance, cleaning, and upgradability. In tubular reactors, the culture flows through unidirectional tubular channels, which allow high liquid flow rates with low shear. In recent times, polyethylene bags are typically used due to their low cost and good light penetration efficiency. Tubular reactors can be several meters long and arranged in different patterns like vertically coiled, horizontally coiled, conical, helical, or 3D mesh layout. A schematic diagram of vertically coiled, horizontally coiled, conical, and helical-type tubular reactors have been shown in Figure 6.1.

Tubular reactors, as reported, require higher construction and installation costs and cumbersome cleaning cycles, are difficult to dismantle and difficult for transportation with additional provision for heat exchange, aeration, and degassing systems. The inclined tubular reactors suffer the loss of energy due to reflection. In present days, cost factors associated with tubular reactors have been nominalised by combining artificial and natural light at the expense of less probability of contamination, high surface area-to-volume ratio, low evaporation losses, high biomass productivity, and high-value product synthesis. Flat-panel reactor (FPR) design provides layer-by-layer arrangement of the reactor with alternate light source arrangement and hence better light penetration inside the culture and better light utilisation efficiency; the vertical arrangement of plates usually save space. Land footprint of FPR as reported is higher than tubular or columnar but lower than open tank systems. FPR has been specifically designed for enhancing light-conversion efficiency due to high surface area-to-volume ratio and has largely been exploited for derivation of microalgae-based products. Special attention is needed about its compositionwhile using wastewater in PBR systems because of the enhanced probability of contamination, which may impart deleterious effects on the production of the value-added compounds. Some other photobioreactor designs in specified application fields are: being torus, trapezoidal, dome, bench type but limited to research scale only not appropriate for commercial exploitation as the scale-up may not be economically feasible.

In open-type categories of the bioreactors, along with raceway pond type, multilayered photo bioreactors are also utilised. In a PBR, the proper spatial distribution of light is vital; it could limit the performance of PBR, especially at the higher growth rate of the biomass (Chang, 2017). Higher growth rate enhances the

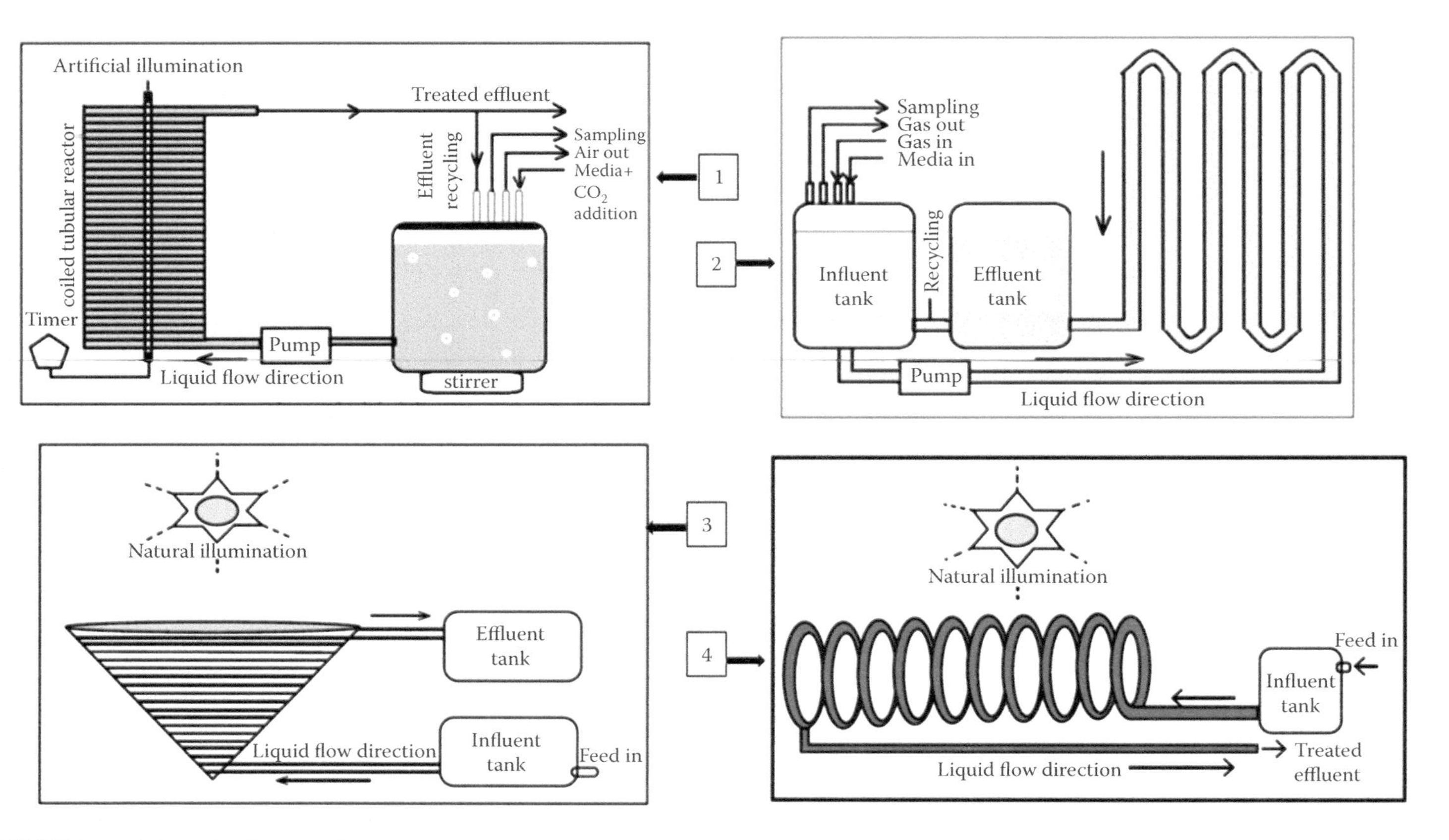

FIGURE 6.1 Schematic diagram of vertically coiled (1), horizontally coiled (2), conical (3), and helical type (4) tubular reactors.

'light shading' effect due to high density of micro-algal cells in a PBR, causing less penetration of light into the culture broth, limiting the photosynthetic ability of cells at the bulk of PBR. Hence, the aim of multilayered PBR design is to enhance light distribution inside the PBR. The structure of the interior compartment of this particular type of PBR being layered enables the dilution and uniform dispersion of incident light in a reactor. In multilayered PBR, microalgal cells are allowed to get immobilised on artificial supports (filter paper pore 0.45 mm), cotton rope, nylon mesh, polystyrene foam, polyethylene screen, to achieve a thin algal biofilm. The partitions used in between the layers are usually arranged by semipermeable membrane layer. Multiple layers of algal biofilms and clear membrane layers remain arranged alternately in an array with gaskets to prevent any sort of leakage and contamination (cell suspension, medium, and products). The semipermeable medium layers permit better light transmittance in every direction. In comparison to conventional plate-type reactors, better light energy dispersion at the interior of the multi-layered PBR, has been reported in several publications (Miron et al., 1999; Razzak et al., 2013).

Some other configurations of PBR of industrial practice are bubble column, air-lift PBRs, and tubular PBRs, which are currently the most widely used closed systems on an industrial scale (Chang,). Several publications have highlighted its potential for wastewater treatment and getting valuable products from that particular type of photobioreactor system. This configuration is highly acceptable for reduced risk of contamination but requires special attention to minimise dead space (stagnant dark interior of the reactor tubes) and maintain proper dimensions of fluid micro-eddies to prevent turbulence associated with damage of cells. In the bubble column and air-lift type configurations, high light transparency and penetration are achieved due to specific design with efficient carbon dioxide bubbling, good mixing with better removal of dissolved oxygen. Air-lift PBRs offers good mixing properties, whereas the configuration of bubble columns allows efficient aeration. Some of the drawbacks of the stirred tank reactors are: use of carbon dioxide bubbles as source of carbon, external positioning of light resources, insufficient provision of light in terms of cost and energy, low surface area-to-volume ratio. Some other specific design features to consider being sparger location, aeration rate (certain minimal level to prevent cell stagnation), gas hold up and liquid velocity, optimal ratio of height to diameter (Mirón et al., 2000; Mirón et al., 2002; Dragone et al., 2010).

Bubble column photobioreactor configurations consist of a cylindrical column with external arrangement of light source with larger surface area-to-volume ratio for good mass and heat transfer, homogeneous mixing facilities and efficient gas flow rate, short light-dark cycle, and enhanced photosynthetic efficiency. The random and inconsistent turbulence by the gas mixture sparging from the bottom of the PBRs may lead to uneven light intensity for a long time and cell sedimentation; hence, scaling up of this design requires proper design of turbulence patterns with increased efficiency by redistribution of coalesced bubbles (Chang,). In air-lift PBRs configurations two interconnecting zones, the riser and the down-comer are arranged, the riser region for gas mixture sparging. The difference in gas holdup between the riser and the down-comer is the main critical parameter in designing air-lift PBRs. Though internal loop, internal loop concentric, and external loop arrangements are available, external loop configuration offers better mixing because of efficient gas disengagement. Air-lift PBRs are characterized by the advantage of generating a circular and homogeneous mixing pattern, specifically controlled residence time of gas in various zones, heat transfer, mass transfer, mixing, and turbulence for micro-algal cultures that are highly fragile and sensitive to shear stress with better biomass yields of broad varieties of microalgae (Monkonsit et al., 2011). The most popular choice among closed PBRs systems are flat-plate PBR, which have a cuboidal shape with a narrow light path, a highly illuminated surface area-to-volume ratio, and open gas transfer area, with both indoor and outdoor culture possibility. The thickness of the plate, surface area-to-volume ratio, and light path length are the decisive factors for process efficiency. However, the thin plates are usually expensive, temperature sensitive, and difficult to clean.

With the aim of improvising the design aspect of the PBR, a newer type of configuration has been practiced and reported as fermenter-type PBRs. This configuration offers open gas exchange, has the ability to monitor/control every operating parameter precisely and accurately, and, presently, is in use for optimization purposes. Industrial scaling-up is still limited because of the low surface area-to-volume ratio and high capital investment. The unique feature of fermenter-type PBR is its suitability with heterotrophic growth of microalgae, for which a light source is not a critical factor. Hang et al. (Chang,) have presented a comparative study on the performance of the different photobioreactors highlighting the advantages and disadvantages of the existing reactor systems. Figure 6.2 presents the selection criteria of a photobioreactor aiming specific application potential.

The design of the photobioreactor needs detailed understanding of the major parameters affecting the PBR operation. Table 6.1 represents the major parameters controlling the operation in PBR.

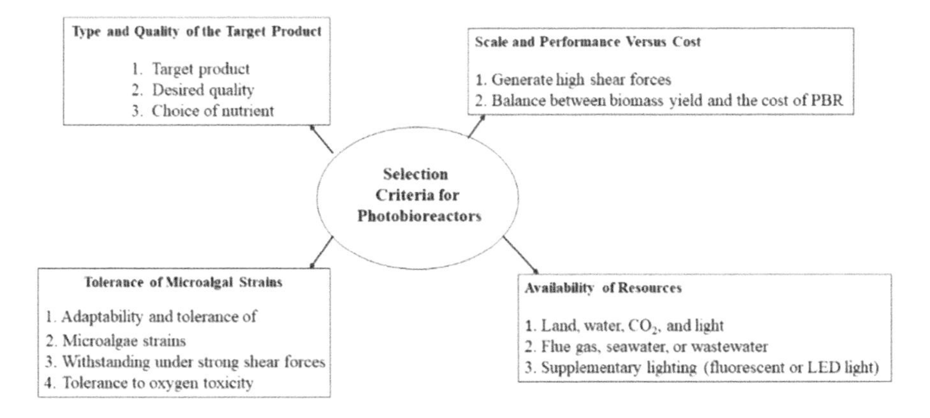

FIGURE 6.2 Selection criteria of a photobioreactor aiming specific application potential.

TABLE 6.1
PBR Operational Parameters for Design Purposes

Operational Parameters	Measurable Parameters
1. Initial nutrient loading rate	1. Measurable parameters. BOD, COD, DO
2. Light requirements • Light intensity • Light duration • Alternate light/dark cycle • Uniformity in light distribution • Distance from illumination source	2. Physiological Parameters • Culture color, odor • Density of culture • Coagulation, flocculation
3. Temperature • Species dependent	3. Temperature, pH, alkalinity, TDS/TSS
4. pH	4. Chlorophyll content
5. Mixing provision • Uniformity in mixing	5. Organic and inorganic carbon
6. Hydraulic retention time (HRT)	6. Lipid content
7. Aeration management • Provision for removing accumulated gases	7. Polysaccharide content
8. Biomass recycling rate	8. Nutrient and metal
9. Regular cleaning and maintenance	9. Biomass, bacterial and algal cell Count
Design Parameters	**Output Parameters**
1. Reactor type • Output dependent • Ease of cleaning, maintenance and scale up	1. Waste water treatment
2. Target population size	2. Metal removal
3. Waste water collection and distribution system	3. CO_2 sequestration
4. Species selection	4. Flue gas sequestration
5. Climatic conditions • Provision of heating and cooling	5. Bio-fuel production
6. Harvesting technology	6. Chemical and protein extraction
7. Economic consideration • Minimal usage of power • Minimal land coverage	7. Bio-hydrogen production
	8. Desulphurization
	9. Bio-fertilizer
	10. Resins and other value added products

6.3 OPERATIONAL FEATURES OF PHOTOBIOREACTORS

The underlying operational principle of the PBR system is to reduce toxicity of the wastewater, i.e. to use the organic load of the wastewater as a nutrient to produce microalgae, which can be utilised as a renewable source of energy. However, the design aspect of the PBR systems sometimes need fine-tuning of the operational and design features for particular waste streams and particular product generation. In the PBR process, nitrogen and organic carbon in wastewater are oxidized by bacteria using oxygen with algal photosynthesis activity, and the produced CO_2 from bacterial activity is consumed by algal cultures. In advanced wastewater treatment plants, with aim for nutrients removal either by nitrification and denitrification, requires a high level of oxygen for nitrification, while denitrification requires an external carbon source. PBR systems can serve the purpose of nutrient removal with simple reactor assembly and valuable product output. PBR systems can overcome the drawbacks of sludge generation and production of greenhouse gas as involved in conventional activated sludge processes and inferior nutrient removal capacity of anaerobic technologies of wastewater treatment. However, the process efficiency of PBR systems are highly dependent on various biotic and abiotic factors (Lu et al., 2020). Light penetration and distribution, the supply of CO_2 as the inorganic carbon source, and nature of waste stream as a nutrient resource are key factors for PBR design dealing with wastewater treatment. The choice of the microorganisms to be grown in PBRs is also an important parameter; phototrophic microalgae are the best choice, being third-generation feedstock for the production of biofuels and bio-based chemicals. Fu et al. (2019) have reported the PBR design with aspect to CO_2 transfer and conversion. This group of authors described the gas-liquid flow pattern in PBR systems and the nutrient utilization strategies. This article has demonstrated that the PBR design strategies are concerned with optimization of gas-liquid two-phase flow and enhancement of light, CO_2, and nutrients transfer. Sayara et al. (2021) have reported the effect of reaction time on nutrients removal from secondary effluent of wastewater. Real field demonstrations were done to assess the removal efficiency of nutrients and organic matter from domestic wastewater. They were carried out using algal-bacterial photobioreactors (Sayara et al., 2021). This group of authors has demonstrated the PBR system dealing with a secondary effluent (domestic discharge after grit removal and sedimentation) form anaerobic pond (Figure 6.3). The process principle involved in the PBR operation was nitrogen removal, which involved primarily the oxidation of ammonium to nitrite, followed by the transfer of ammonium and nitrite mixture into the gravel layer by upward movement with enhanced turbulence, and finally, dual action of algal and bacterial cultures to remove the different forms of nitrogen. This group of authors has claimed efficient nutrient removal from wastewater for real effluent with reaction time as a key system controlling parameter and maximum COD removal rates of 50% (under RT of 4 h). No significant effect on introducing nitrifiers was recorded on the removal of both COD and phosphorus level; 95% removal of the phosphorus was achieved and ammonium ($NH4^+$) removal rate was found enhanced significantly with the addition of nitrifiers.

Similar studies related to the removal of nitrogen and phosphorus from waste effluent using the PBR approach have also been reported by several other groups of authors (Muñoz and Guieysse, 2006; Kim et al., 2013; Babaei et al., 2016; Delgadillo-Mirqueza et al., 2016; Shayan et al., 2016; Nordlander et al., 2017; Tang et al., 2018; Rada-Ariza et al., 2019; Shahid et al., 2020). Tang et al. (1997) have presented a detailed study on the choice of the bacterial consortium for tertiary wastewater treatment at cold environment, polar cyanobacteria being the best

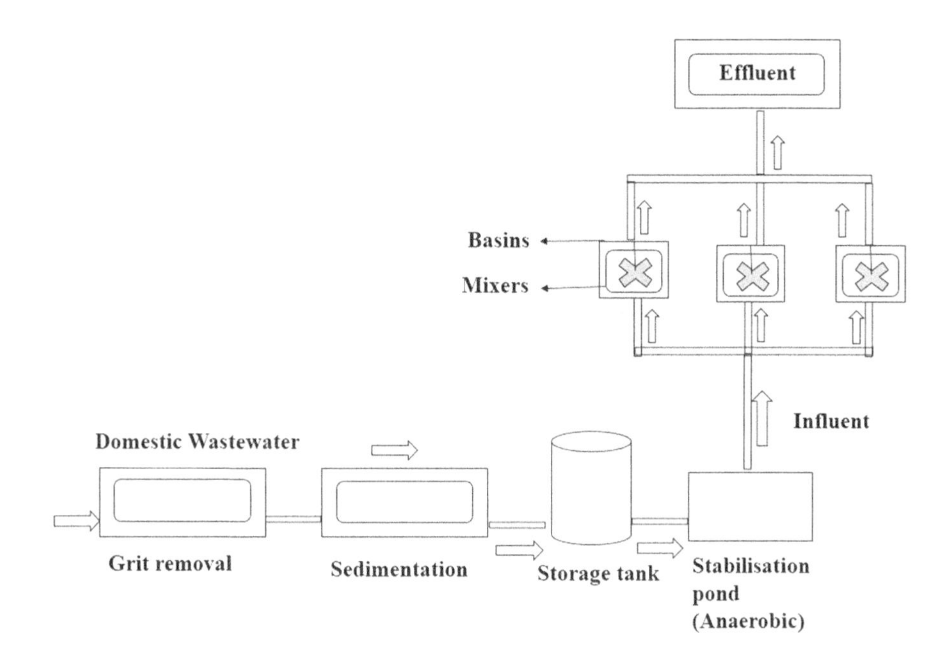

FIGURE 6.3 Schematic of PBR system dealing with a secondary effluent from domestic discharge.

choice for cold environments, while the green algae were chosen for summer conditions. Ashok et al. (2019) have demonstrated the performance of algal-bacterial photobioreactor, highlighting the effect of nutrient load and light intensity. This group of author-illustrated nutrient removal efficiency from secondary treated domestic wastewater used algal-bacterial semibatch photo bioreactors, with the main focus on optimization of illumination costs and nutrient removal efficiency to offer an economically viable PBR system. The algal bacterial combination used for the study was Chlorella *vulgaris* and C*hlamydomonas reinhardtii* with detailed exploration on biomass yield, dissolved organic carbon, and chlorophyll content. A maximum of ~25 mg l^{-1} N and ~10 mg l^{-1} P were reported to be efficiently removed with HRT of 2 days, 9:3 light: dark condition and 1,543 μmol photons m^2 s^{-1} light intensity. Praveen and Loh (2016) (Sharma et al., 2016) have reported a novel type of PBR system for nitrogen and phosphorus removal from tertiary wastewater in an osmotic membrane photobioreactor. As reported, application of the photobioreactors for wastewater treatment was not only limited to the nutrient removal but also had been practiced for several other specific applications, like removal of pharmaceutical contaminants and associated derivatives. As per WHO, emergence of antibiotic resistance has become a global health concern of the 21st century arising out the severe contamination of municipal, agricultural, and hospital discharge to water bodies (Sharma et al., 2016). Wastewater treatment plants receiving municipal wastewater suffers from treatment complexity arising from antibiotic resistance genes (ARGs) in natural environments (Nõlvak et al., 2018). The study of Nõlvak et al. (2018) has focused on the fate of 'antibiotic resistome' and 'integrin-integrase genes (class 1–3)' during treatment in photobioreactors using several bacterial genera, like *Azoarcus*, *Dechloromonas*, and *Sulfuritalea*. Their approach of exploiting photobioreactors has been proven effective in controlling the antibiotic-resistance propagation effect for municipal wastewater as compared to conventional wastewater treatment approaches. The reduction of resistome was reported to be affected by nature of the bacterial-algal consortium. In their study, the multidrug-resistance approach was not reported but has guided a new approach of treating drug contamination coupled with technical and methodological details highlighting the need for further in-depth investigation. The additional investigation would give a comprehensive view on this critical subject area with technological innovations. Zaineb et al. (2019) have reported the use of a photobioreactor for treatment of textile effluent focusing on decolorization and phytotoxicity reduction. Typical textile effluents are characterized by a complex composition of organic matter, high turbidity, extreme (acidic/alkaline) pH, and high color concentration due to the poor solubilization of dyes. Discharge of chromophoric compounds in water bodies may reduce light penetration, hindering photosynthesis and causing reduction in DO levels in natural ecosystems. The environmental sustainability limitations associated with the physical and chemical processes of textile effluent treatment due to generation of secondary pollutants has led the research and development of the environmental sustainable PBR technologies. A novel type anaerobic/aerobic algal-bacterial photobioreactor for the treatment of synthetic textile wastewater has been attempted emphasizing organic carbon, nitrogen, and phosphorous removal efficiencies based on azo and anthraquinone dyes. As stated by this group of authors, algal-bacterial symbiosis may be a cost-competitive alternative for the treatment of wastewaters with a low C/N ratio as the symbiotic interaction is based on the mutualistic exchange of O_2 and CO_2 between microalgae and bacteria. It has been reported that the microalgae cell walls has facilitated dye mineralization by pollutant adsorption, while bacteria has improved microalga growth by realizing growth-promoting factors. It has also been reported that the treated water exhibited no inhibition over seed germination, while the raw textile effluent exhibited about 52% inhibition. Praveen and Loh (2016) have reported a novel type PBR system for nitrogen and phosphorus removal from tertiary wastewater in an osmotic membrane photobioreactor. A membrane photobioreactor for tertiary sewage treatment has also been reported by González-Camejo et al. (2019). Membrane photobioreactor (MPBR) technology (combination of membrane and microalgae cultivation) has been classified as a novel water-resource recovery facility avoiding huge energy demands and nutrient losses associated with the conventional wastewater treatment plants. The integration of membrane filtration with PBRs helps to achieve high-quality permeate, free of suspended solids and pathogens, and hence, may act as source of reclaimed water. The filtration of microalgae through membrane was reported to offer shorter hydraulic retention times (HRTs) and longer biomass retention times (BRTs), recovery of large quantities of nutrients without purging of microalgae culture. This group of authors has also reported enhancement of microalgae performance, enhanced nutrient load of system, thereby reducing the large land requirement for microalgae cultivation. Studies of

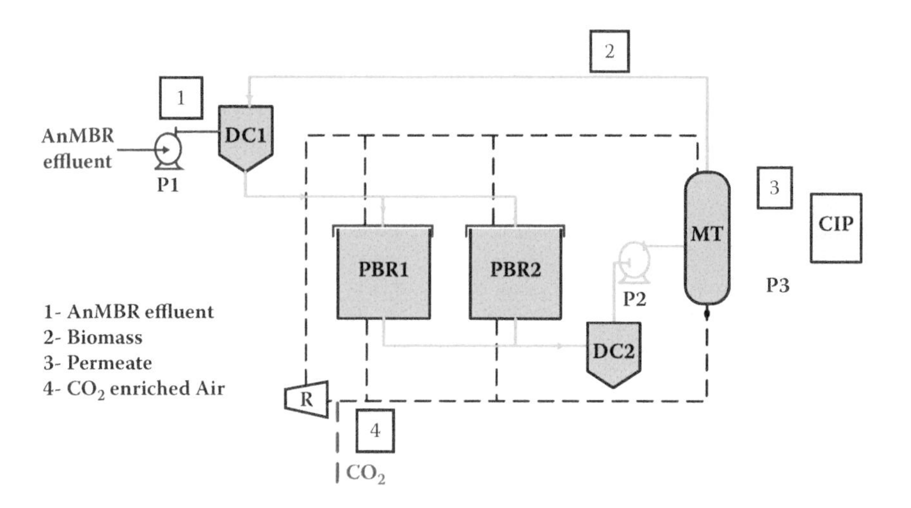

FIGURE 6.4 Process schematic of outdoor MPBR plant dealing with coal industry effluent.

González-Camejo et al. (2019) have reported that in outdoor conditions 3.8-fold higher nitrogen and phosphorus recovery and double biomass productivity in a membrane PBR in comparison with simple PBR system with significant lower land requirement. The process schematic of the outdoor plant MPBR plant has been represented in Figure 6.4. This group of authors has also reported the effect of critical design factors like culture recirculation mode, the non-photic volume of the MPBR plant (dark zone), BRT and HRT, membrane fouling and temperature variation. Final treated water was reported to have negative response for pathogens like *E.coli* suggesting a promising source of reclaimed water suitable for irrigation or different urban and industrial purposes.

Baral et al. (2020) have reported the potential of wastewater treatment processes through algal PBR systems as a biofuel production process. The process was integration of an anaerobic membrane reactor and algal photoreactor with utilization of sewage treatment. Plant design was done to convert sewage COD to methane and carbon dioxide, while the second step involved utilization of the CO_2 for algal growth using sunlight as the energy source. Finally, it involved production of the biofuel from cultured microalgae. The system was a illustration of sustainable process aiming at complete conversion of total carbon (TC) in the wastewater into energy and zero discharge plant. A study of Luo et al. (2021) has reported the in-depth analysis and resilience of photobioreactors for hazardous wastes (chemical shocks, including sodium chlorite, sodium hypochlorite, and 2,4-dinitrophenol). Authors have demonstrated the deleterious effect of chemicals on photobioreactors performance, which was acclimated to domestic wastewater influent and affirmed its capability to cope with low levels of chemical stress under optimized conditions. PBR technology has also been reported for effective wastewater treatment of coal-propelled power plants. The study of Cohena et al. (2020) has demonstrated a unique PBR system having the potential to be used in large-scale domestic and industrial wastewater treatment plants. The novel PBR system interior was provided with multiple high-porous polypropylene beds, enabling the system adaptable for pollutants both from gaseous and aquatic effluents. The design was based on wastewater treatment plant habitat-endemic flora (single cyanobacteria *Synechocystis salina*) not on preserved culture strain. Authors have claimed that the photosynthetic biomass in that novel PBR system preferably grew on PP beds and have effectively assimilated the biomass nutrients, such as phosphorous, nitrogen, boron, magnesium, calcium, and the toxic ions like strontium and fluoride along with conventional PBR performance of mitigating greenhouse gas emission by the assimilating bicarbonate (HCO_3^-) from the aquatic phase and CO_2 from the gas phase. The choice of microbial culture *Synechocystis salina* was based on its resilience to high salinity being active at very low concentrations of essential nutrients and low N: P ratio of 0.1–0.2. During the algal blooming periods, maximal Sr^{2+} removal was reported 74%, maximal HCO_3^- removal 91%, and total nitrogen (TN) removal 98%. A study of García-Galán et al. (2021) has illustrated a semi-closed, tubular, horizontal photobioreactor (PBR) for treatment of irrigation: rural drainage water mixture, focusing on the removal of different contaminants (pharmaceutical component, personal care products and flame retardants), and evaluation of the environmental impact of the resulting effluent. The study was based on treatment of six drug components, two flame retardants, three fragrances, and pesticide components. As reported, treated effluent with hazard quotients (HQs) values < 0.1 (no risk associated) was obtained for most of the trophic level concerned (like green algae, aquatic animals, and invertebrates). Significantly high elimination efficiencies were obtained for the benzodiazepines, oxazepam (highly recalcitrant), and diazepam, which are rarely removed by conventional treatment systems. Hence, based on the comprehensive review of the prospect and utilization pattern of the PBR systems, it may be concluded that PBRs are a promising tool for wastewater remediation in which many benefits may be gained along with a diverse application potential for various waste effluent treatment in a more controlled manner than in natural ecosystems. However, several industrial applications of PBR systems are still in practice, but considering the optimum balance between economic feasibility and environmental sustainability, design aspects of PBR systems still need exploration in terms of material of construction, illumination type, and dispersion pattern, temperature variation, type of waste treatment (nutrient type), nature of the culture, and discharge quality.

6.4 CONCLUSION AND FUTURE PROSPECTS

Photobioreactor systems though offers several advantages while coupled with wastewater treatment schemes, like i) removal of a plethora of harmful materials, such as heavy metals, petroleum hydrocarbons, pesticides, pharmaceuticals, nanomaterials, and personal-care products, ii) production of low-cost biomass for methane and biofuel-generation organic fertilizer, and other valuable products, iii) recovery of valuable nutrients and heavy metals removal, iv) advanced cost-effective and low-energy wastewater treatment facility, as compared to conventional biological nutrient removal schemes dealing with high energy processes, v) treatment of phosphorous in the effluents, lowering the probability of eutrophication for further application in agricultural sectors, but still the technology is not fully established to a full extent because of the limitations of appropriate design aspects and associated cost factors. In present status, the standalone PBRs did not meet the objectives of renewable energy and economic targets. There are ample further studies on PBR systems in diverse directions, like (i) further research on role of biomass and its choice for adsorptive removal of critically emerging contaminants contributing to establish complete mass balances in microalgae systems and detail biomass analysis, (ii) optimization of potentially influencing parameters, such as temperature, pH, flow characteristics, hydraulic retention time, nature of living organisms with process performance, (iii) modifications of the existing PBR configurations, like design of inclined algal biofilm photobioreactor, providing baffles or static mixers inside flat or column PBRs to enhance mixing, and promotion of more frequent passage of cells under the light source, (iv) design of enclosed nonconventional PBR systems, like curved-chamber PBR, V-shaped PBR, alveolar panel PBR, tilted-rocking flat-plate PBR, dome-shaped PBR, vertically-stacked horizontal PBR, etc. (Assunção and Xavier Malcata, 2020). In present perspective, a number of improvements need to be incorporated in classical configurations, considering all probable advantages along with plausible detriments, to address the major challenges associated with scalability, operating costs, and configuration issues. There is not any universally ideal PBR design. The best design is yet to come, based on utilization criteria.

REFERENCES

Ashok V., Gupta S.K., and Shriwastav A. (2019). Photobioreactors for wastewater treatment. In Gupta, S.K., and Bux, F. (Eds.), *Application of microalgae in wastewater treatment* (pp. 383–409. Springer, Cham. doi:10.1007/978-3-030-13913-1_18.

Ashok, V., Shriwastav, A., Bose, P., and Gupta, S.K. (2019), Phytoremediation of wastewater using algal-bacterial photobioreactor: Effect of nutrient load and light intensity. *Bioresouce Technol. Rep.*, *2*, 100205.

Assunção, J., and Xavier Malcata, F. (2020). Enclosed "non-conventional" photobioreactors for microalga production: A review. *Algal Research*, *52*, 102107.

Babaei, A., Mehrnia, M.R., Shayegan, J., and Sarrafzadeh, M.H. (2016). Comparison of different trophic cultivations in microalgal membrane bioreactor containing N-riched wastewater for simultaneous nutrient removal and biomass production. *Process Biochem.*, *51*, 1568–1575.

Baral, S.S., Dionisi, D., Maarisetty, D., Gandhi, A., Kothari, A., Gupta, G., and Jain, P. (2020). Biofuel production potential from wastewater in India by integrating anaerobic membrane reactor with algal Photobioreactor. *Biomass Bioenergy*, *133*, 105445.

Biel-Maeso, M., Corada-Fernández, C., and Lara-Martín, P.A. (2018). Monitoring the occurrence of pharmaceuticals in soils irrigated with reclaimed wastewater. *Environ. Pollut.*, *235*, 312–321. doi:10.1016/j.envpol.2017.12.085.

Chang, J.-S., Show, P.-L., Ling, T.-C., Chen, C.-Y., Ho, S.-H., Tan, C.-H., Nagarajan, D., and Phong, W.-N. Current Developments in Biotechnology and Bioengineering: Bioprocesses, Bioreactors and Controls. Elsevier. doi:10.1016/B978-0-444-63663-8.00011-2.

Cohena, Y., Nisnevitcha, M., and Ankera, Y. (2020). Innovative large-scale photobioreactor for coal propelled power plant, effluents treatment. *Algal Research*, *52*, 102101.

Delgadillo-Mirqueza, L., Lopes, F., Taidi, B., and Pareau, D. (2016). Nitrogen and phosphate removal from wastewater with a mixed microalgae and bacteria culture. *Biotechnol. Rep.*, *11*, 18–26.

Dragone, G., Fernandes, B.D., Vicente, A.A., and Teixeira, J.A. (2010). Third generation biofuels from microalgae. In Me´ndez-Vilas, A. (Ed.), *Current research, technology and education topics in applied microbiology and microbial biotechnology* (pp. 1255–1366). Formatex Research Center, Badajoz, Spain.

Fu, J., Huang, Y., Liao, Q., Xia, A., Fu, Q., and Zhu, X. (2019). Photo-bioreactor design for microalgae: A review from the aspect of CO_2 transfer and conversion. *Bioresource Technology*, *292*, 121947.

García-Galán, M.J., Matamoros, V., Uggetti, E., Díez-Montero, R., and García, J. (2021). Removal and environmental risk assessment of contaminants of emerging concern from irrigation waters in a semi-closed microalgae Photobioreactor. *Environ. Res.*, *194*, 110278.

González-Camejo, J., Jiménez-Benítez, A., Ruano, M.V., Robles, A. Barat, R., and Ferrer, J. (2019). Optimising an outdoor membrane photobioreactor for tertiary sewage Treatment. *J. Environ. Manage.*, *245*, 76–85.

Huang, Q., Jiang, F., Wang L., and Yang, C. (2017). Design of Photobioreactors for mass cultivation of photosynthetic organisms. *Engineering*, *3*, 318–329.

Jesús García-Galán, M., Matamoros, V., Uggetti, E., Díez-Montero, R., and García, J. (2021). Removal and environmental risk assessment of contaminants of emerging concern from irrigation waters in a semi-closed microalgae Photobioreactor. *Environ. Res.*, *194*, 110278.

Kim, J., Liu, Z., Lee, J.Y., and Lu, T. (2013). Removal of nitrogen and phosphorus from municipal wastewater effluent using Chlorella vulgaris and its growth kinetics. *Desalin. Water. Treat.*, *51*, 7800–7806.

Lu, W., Alam, M.A., Liu, S., Xu, J., and Saldivar, R.P. (2020). Critical processes and variables in microalgae biomass production coupled with bioremediation of nutrients and CO2 from livestock farms: A review. *Sci. Total Environ*, *716*, 135247.

Luo, Y., Logan, A., Henderson, R.K., and Le-Clech, P. (2021). Evaluating the resilience of photobioreactors in response to hazardous chemicals. *Chem. Eng. J.*, *405*, 126666.

Mamo, J., García-Galán, M., Stefani, J.M., Rodríguez-Mozaz, S., Barceló, D., Monclús, H., Rodriguez-Roda, I., and Comas, J. (2018). Fate of pharmaceuticals and their transformation products in integrated membrane systems for wastewater reclamation. *Chem. Eng. J.*, *33*, 450–461. doi:10.1016/j.cej.2017.08.050.

Miron, A.S., Gomez, A.C., Camacho, F.G., Grima, E.M., and Chisti, Y. (1999). Comparative evaluation of compact photobioreactors for large-scale monoculture of microalgae. *Journal of Biotechnology*, *70*, 249–270.

Mirón, A.S., Camacho, F.G., Gómez, A.C., Grima, E.M., and Chisti, Y. (2000). Bubble-column and airlift photobioreactors for algal culture. *Bioengineering, Food, and Natural Products*, 46, 1872–1887.

Mirón, A.S., Garcıa, M.-C.C., Camacho, F.G., Grima, E.M., and Chisti, Y. (2002). Growth and biochemical characterization of microalgal biomass produced in bubble column and airlift photobioreactors: Studies in fed-batch culture. *Enzyme Microb. Technol.*, *31*, 1015–1023.

Monkonsit, S., Powtongsook, S., and Pavasant, P. (2011). Comparison between airlift photobioreactor and bubble column for Skeletonema costatum cultivation. *Engineering Journal*, *15*, 53–64.

Muñoz, R., and Guieysse, B. (2006). Algal-bacterial processes for the treatment of hazardous contaminants, A review. *Water Res.*, *40*, 2799–2815.

Ngo, HH., Hoang, N., Phong, N., Wenshan, V., Xuan-Thanh, G., Phuoc Dan, B., Thi, N., Hong, M., and Xinbo Zhang, N. (2018). Advances of photobioreactors in wastewater treatment: Engineering aspects, applications and future perspectives. In Bui, X.T., Chiemchaisri, C., Fujioka, T., and Varjani, S. (Eds.), *Water and wastewater treatment technologies. Energy, environment, and sustainability*. Springer, Singapore (pp. 297–328). doi:10.1007/978-981-13-3259-3_14.

Nhat H., Vo, P., Hao Ngo, H., Guo, W., Hong Nguyen T.M., Liu,Y., Liu, Y., Duc Nguyen, D., and Woong Chang, S. (2019). A critical review on designs and applications of microalgae-based photobioreactors for pollutants treatment. *Science of the Total Environment, 651*, 1549–1568.

Nordlander, E., Olsson, J., Thorin, E., and Nehrenheim, E. (2017). Simulation of energy balance and carbon dioxide emission for microalgae introduction in wastewater treatment plants. *Algal Res., 24*(Part A), 251–260.

Nõlvak, H., Truu M., Oopkaup, K., Kanger, K. Krustok I., Nehrenheim, E., and Truu, J. (2018). Reduction of antibiotic resistome and integron-integrase genes in laboratory-scale photobioreactors treating municipal wastewater. *Water Res., 142*, 363–372.

Palmer, C.M. (1974). Algae in American sewage stabilization's ponds. *Rev Microbiol (S-Paulo), 5*, 75–80.

Praveen, P., and Loh, K.C. (2016). Nitrogen and phosphorus removal from tertiary wastewater in an osmotic membrane photobioreactor. *Bioresour. Technol., 206*, 180–187.

Racar, M., Dolar, D., Karadakić, K., Čavarović, N., Glumac, N., Ašperger, D., and Košutić, K. (2020). Challenges of municipal wastewater reclamation for irrigation by MBR and NF/RO: physico-chemical and microbiological parameters, and emerging contaminants. *Sci. Total Environ., 722*, 137959.

Rada-Ariza, A., Fredy, D., Vazquez, C.L., van der Steen, P., and Lens, P. (2019). Ammonium removal mechanisms in a microalgal-bacterial sequencing-batch photobioreactor at different solids retention times. *Algal Res., 39*, 101468.

Razzak, S.A., Hossain, M.M., Lucky, R.A., Bassi, A.S., and Lasa, H. de (2013). Integrated CO_2 capture, wastewater treatment and biofuel production by microalgae culturing-A review. *Renewable and Sustainable Energy Reviews, 27*, 622–653.

Rueda, E., García-Gal´an, M.J., Díez-Montero, R., Vila, J., Grifoll, M., and García, J., (2020). Polyhydroxybutyrate and glycogen production in photobioreactors inoculated with wastewater borne cyanobacteria monocultures. *Bioresour. Technol., 295*, 122233.

Sayara, T., Khayat, S., Saleh, J., and Van Der Steen, P. (2021). Evaluation of the effect of reaction time on nutrients removal from secondary effluent of wastewater: Field demonstrations using algal-bacterial photobioreactors. *Saudi J. Biol. Sci., 28*, 504–511.

Shahid, A., Malik, S., Zhu, H., Xu, J., Nawaz, M.Z., Nawaz, S., Alam, M.A., and Mehmood, M.A. (2020). Cultivating microalgae in wastewater for biomass production, pollutant removal, and atmospheric carbon mitigation; A review. *Sci. Total Environ., 704*, 135303.

Sharma, V.K., Johnson, N., Cizmas, L., McDonald, T.J., and Kim, H. (2016). A review of the influence of treatment strategies on antibiotic resistant bacteria and antibiotic resistance genes. *Chemosphere, 150*, 702–714.

Shayan, S.I., Agblevor, F.A., Bertin, L., and Sims, R. (2016). Reaction time effects on wastewater nutrient removal and bioproduct production via rotating algal biofilm reactor. *Bioresour. Technol., 211*, 527–533.

Sun, Y., Huang, Y., Liao, Q., Fu, Q., and Zhu, X. (2016). Enhancement of microalgae production by embedding hollow light guides to a flat-plate photobioreactor. *Bioresour. Technol., 207*, 31–38.

Sun, Y., Huang, Y., Liao, Q., Xia, A., Fu, Q., Zhu, X., and Fu, J. (2018). Boosting Nannochloropsis oculata growth and lipid accumulation in a lab-scale open raceway pond characterized by improved light distributions employing built-in planar waveguide modules. *Bioresour. Technol., 249*, 880–889.

Tang, C.C., Tian, Y., He, Z.W., Zuo, W., and Zhang, J. (2018). Performance and mechanism of a novel algal-bacterial symbiosis system based on sequencing batch suspended biofilm reactor treating domestic wastewater. *Bioresour. Technol., 265*, 422–431.

Tang, E.P.Y., Vincent, W.F., Proulx, D., Lessard, P., and Noüe, J.d. (1997). Polar cyanobacteria versus green algae for tertiary waste-water treatment in cool climates. *J. Appl. Phycol.*, *9*, 371–381.

Zaineb, D., Toledo-Cervantes, A., Ghedira, K., Chekir-Ghedira, L., and Muñoz, R. (2019). Decolorization and phytotoxicity reduction in an innovative anaerobic/aerobic photobioreactor treating textile wastewater. *Chemosphere*, *234*, 356–364.

Zhu, L. (2015). Biorefinery as a promising approach to promote microalgae industry: An innovative framework. *Renew. Sustain. Energy Rev.*, *41*, 1376–1384.

7 Genetic Engineering of Algae

Sivakumar Durairaj and Perumalla Srikanth
Kalasalingam School of Agriculture and Horticulture, Kalasalingam Academy of Research and Education, Anand Nagar, Krishankoil, Tamil Nadu, India

Shankar Durairaj
K.M. College of Pharmacy, Uthangudi, Madurai, Tamil Nadu, India

CONTENTS

7.1 INTRODUCTION

Food production is expected toare to be identified through be augmented to meet the needs of 9.7 billion people worldwide by 2050, according to FAO (Ng et al., 2017). At present, people use around half of petrol subsidiaries, depleting resources over the long run. The consumption of nonenvironmentally friendly power sources has

DOI: 10.1201/9781003165101-7

incited an addition in CO_2 levels in the climate, provoking an overall temperature adjustment and disturbing climatic parts (Behera et al., 2015). Additional sources, improvements over the existing ones, will be explored to meet the demands in science and development. Microalgae have emerged as one of the food supplement sources in the past few years (Chia et al., 2018). Microalgae also help make proteins, color pigments, biofuels, nutraceuticals, drugs, and other value-added products (Fayyaz et al., 2020).

Microalgae are from the arranged eukaryotic lineages and are progressed through different basic endosymbiosis actions (Nelson, 2017). Generally, assortment metabolic pathways and authoritative associations stay unseen owing to genetic assortment and alterability. With the shortfall of metabolic pathways, it is very difficult to work with some of the vulnerable algae in algal bioengineering. A few metabolic genome-based models were opened for microalgae (Koussa et al., 2014) and to sequence the genomes of high-quality species. Regardless of genome-changing algae advances, they are consistently involved in human cell lines and enhance the attainment in microalgae with low zeroing in on adequacy. This article depicts a couple of resources, like the *Chlamydomonas* crack collection library, and highlights the continuous movements by genome modifying and algal bioengineering, which result in photosynthetic benefits and the creation of huge value-added products (Fu et al., 2019).

Around 44,000 species have been legitimately portrayed, yet the certified number, subject to how "green development" is used, will be significantly higher; specific assessments are comparably high for a million species. As photoautotroph natural elements, green development is the early phase of most food networks in land and water-proficient conditions. The biomass effectiveness of various algal species has diverged from that of natural plants, which can beneficially evolve without the intervention of contamination specialists and pesticides in new or sea water. These factors, similar to the proliferation of supplements, polyunsaturated unsaturated fats, and other sound blends, have incited extending client premium and business interest in green development creation during the last decades. Algae are a polyphyletic social occasion of animals from four unmistakable natural domains: bacteria, plantae, chromista, and protozoa. Around 44,000 species have been consistently portrayed, with the verified number dependent on which significance of "green development" is used; it could be significantly higher, with appraisals being as high as 1,000,000 particular species (Ullmann and Grimm, 2019).

As photoautotroph animals, green development is the early phase of most food networks in maritime conditions. The biomass productivity of various algal species is significantly higher compared to that of natural plants, and they can be capably evolved without microbial and pesticides in new or sea water. These components, similar to their high substance of supplements, polyunsaturated unsaturated fats, and other sound blends, have incited new uses of microalgae, including diatoms, green events, and cyanobacteria, have been scrutinized for over a century. The utilization of these photosynthetic living creatures provides a few benefits for overdeveloped plants concerning headway rate, higher biomass profitability, non-appearance of conflict with crops in arable regions, and the limit concerning reasonable improvement using periphery assets (Levering, 2016).

The production of genetically modified organisms (GMOs) is tangled. The term GMOs can encompass everything seen as risky to people and animals, considering potential consequences of undesired quality streams in the climate. Consequently, algomics is valuable in masterminding tests and expecting conceivable joint exertion. It aids in considering metabolic pathways, impulses, and genome groupings, and subsequently helps metabolic and natural arrangements. Inborn changes on various microalgae species were performed with unequivocal strains that are showing acquired power and unreliability to hereditary change. It breaks a test for strength control in the hereditary arrangement of microalgae. Despite utilizing CRISPR improvement for focused genome changing in microalgae, it produces fiscally real microalgae strains. More focus should be given to the human-affirmed strains since any damaging straightforwardness of brand-name strains could affect the whole environment and results. Study and danger valuation should occur before passing on GMOs from the lab to typical habitats (Fu et al., 2019).

Riboswitches, a novel instrument, were used by the compelling verbalization of transgenes for making quality rules; they link nuclear organizers, luciferase for the surge characteristics, and chloroplast quality verbalization. Either a nuclear apparatus stash or inherited gadget information is necessary to discover the metabolites and interpret metabolic pathways. The examination of the game plan will recall the data for metabolomics that opens the domain for the metabolic change assessment through metabolic associations. The green development of single cells are adequately tillable and locked in metabolic planning (Fajardo et al., 2018).

Change in the model microalga *Chlamydomonas reinhardtii* was refined by the change methodologies used for plants; it was cultivated for both nuclear genomes and chloroplast genomes. The changes by the glass globule's unsettling influence were also especially made for *C. reinhardtii*, where protoplast cells were vexed energetically in glass spots. All things considered, movement of inherited material in microalgae is seen as significantly less efficient than movement concerning plants. Once transformants are procured, microalgae are generally single cells that can be taken care of without requiring the repetitive recuperation stages in plants (Jeon et al., 2017).

Moreover, RNAi advancement can similarly be used to reduce control-quality verbalization, especially for the light-procuring radio-wire buildings using LHC proteins in *Chlamydomonas reinhardtii*. The planned light-social occasion strains have the properties of high photo resistivity and the light passage limit. Likewise, the period from initial transformation, the creation stage, is reduced, appearing differently from mammalian-based stages. *C. reinhardtii* is one of the good resources for the generation of unsaturated fats and hydrogen (Anand et al., 2017). Current advancements engaged for characteristics control reflect the structure-based limit and other sub-nuclear applications (Anand et al., 2017; Fajardo et al., 2018).

Diatom's advance has experienced the presence of productive genome-evolving frameworks, which don't empower the utilitarian evaluations, paying little attention to the speedy improvement of the synopsis of oleaginous microalgae genomes sequencing. The advancement in the genomes is (i) increasing the interest in studying microalgae science, and (ii) improving the strain in biotechnological applications. At this point, it is conceivable that post-transcriptional down-instruction would be valuable for articulation (RNAi) or employing lines with over-stated qualities in diatoms. These days, facilitated genome changing is possible, at least for a couple of microalgae animal social events.

The genome change has been created by homologous recombination nitrite reductase, ferritin in *Nannochloropsis*, *Ostreococcus* utilizing zinc-finger nucleases and changing neighborhood attributes of green microalgae known as *C. reinhardtii*. The transcriptional activator-like effector nucleases (TALENs) were used in mutagenesis of *Phaeodactylum tricornutum* (Weyman et al., 2015). Furthermore, the CRISPR / CRISPR-Activated structure is used for the presentation of constant quality changes in the green microalgae *C. reinhardtii* (Shin et al., 2016) and *P. tricornutum* (Nymark et al., 2016).

7.2 ALTERATION APPARATUS FOR GM MICROALGAE DEVELOPMENT

The successful stage of movement of microalgae for planned bioproducts requires an gene mechanical (GM) assembly compartment. Having an arrangement of checked mind-boggling contraptions will empower control of the algal genome to pass on ideal proportions of the objective things. Different microorganisms, such as minuscule living things and yeast, have evolved to influence a changing course in current things using the wide-arranging gadget sets (Gopal and Kumar, 2013). Microalgae are acquiring energy to be commensurate to these other mechanical microorganisms, with ceaseless advances in the new events and improvement of new natural sections and change measures.

In addition to typical things, nuclear and genetic gadgets offer astonishing approaches to improve the formation of centered particles. The field lacks the progression of a cutting-edge setup of inherited instruments for microalgae; it appears different from other mechanical microorganism stages. In light of everything, the toolsets now available support the speed and exactness with which one can configure green development to progress effectively. Metabolic planning involves focus for express metabolic pathways inside cells to change the progress of metabolites near an ideal thing (Naghshbandi et al., 2020). Alteration apparatus for GM microalgae development is shown in Figure 7.1.

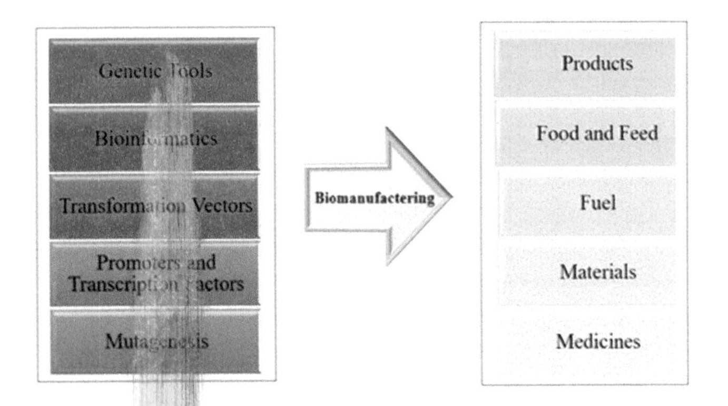

FIGURE 7.1 Alteration apparatus for GM microalgae development.

7.2.1 Arrangement of Enunciation Vectors

Designed science incorporates the arrangement of enunciation vectors with demonstrated inherited segments for achieving an ideal total or limit in the model animal. Splendid gate modular cloning (MoClo) device stashing is currently available for most critical microorganisms, which grant vectors periods by giving a collection of inherited design blocks for arrangement and advancement access (Martella et al., 2017). Lately, Crozet et al. (2018) found the important MoClo apparatus compartment for *Chlamydomonas reinhardtii,* which contains more than 100 quality parts with 67 remarkable innate segments. The UTRs, eliminators, names, columnists, antidote poison-resistance characteristics, and introns are a homogenous collection of confined parts for making codon-updated verbalization vectors, which provide a great explanation in *C. reinhardtii*. An affirmed instrument stash was made available for the assessment of neighborhood license advanced planning of *Chlamydomonas.* The availability of new innovation enhances the algal biotechnology sector by lessening progression time and providing solitary assessment social events. Besides, the fundamental high cutoff quality of stacking device compartments for *Nannochloropsis* was arranged in a nearly detached association, including distinctive flexible promoters, deterrent indicators, and writer characteristics, into gateway-suitable vectors; these join in a suitable plan for reaching multi-quality enunciation. Achieving multi-quality verbalization is fundamental for empowering pathway planning, where different builds are locked in with the biosynthesis of significances of responsiveness. The new quality loading structure headway will permit really obvious and profitable metabolic-planning cycles. In addition, another bioluminescent writer system for the assessment of value enunciation was found for *Nannochloropsis.* These firefly luciferase writers are used to verify designed circadian rhythms, and moreover, enhance to the broad nuclear device compartment that has been made for this species (Poliner et al., 2019).

7.2.2 Transgenes Encoding

To achieve critical returns of recombinant proteins and metabolic things, it is essential to recognize cis-regulatory segments prepared for driving ideal explanations of transgenes encoding for pertinent consequences. Publicists who dependably provoked top-notch enunciation have highlighted countless noteworthy algal strains, like fanciful HSP70/rbcs2 sponsor (typically suggested as 1) from *C. reinhardtii* and the diatom *Phaeodactylum tricornutum.* These promoters play a significant role while making multi-quality forms to evade the impacts of using a comparable sponsor on various events in a creation, which can achieve innate strategies after some time.

Furthermore, while using metabolic pathway quality stacking, transgenes frequently must be conveyed on differentiating levels to track down some sort of amicability required for most noteworthy yields (Ramarajan et al., 2019). Continuous exposures of innovative endogenic sponsors, including constitutive and inducible limits, diatoms, and cyanobacteria, are presented in Table 7.1. While fundamental supporters take advantage of making low-support algae transformants,

TABLE 7.1
List of Different Promoters Used in GM Algae Development

Name of the Promotor	Abbreviation	Origin Organism	Function	Product	References
Extrinsic protein in photosystem II	EPPSII	*Nannochloropsis gaditana*	Constitutive expression of transgenes 4.5 × higher than β-tubulin promoter	mVenus reporter	Ramarajan et al., 2019
Heat shock protein 70	HSP90	*Nannochloropsis gaditana*	Constitutive expression of transgenes 3.1 × higher than β-tubulin promoter	mVenus reporter	Ramarajan et al., 2019
PhotosystemI protein D	CvpsaD	*Chlorella vulgaris*	Weak constitutive expression of transgenes	Aminoglycoside 3′ -phosphotransferase (aphVIII)	Kim et al., 2018
Glycerol-3-phosphate dehydrogenase	CrGPDH3	*Chlamydomonas reinhardtii*	Strong salt-inducible overexpression of transgenes	GUSPlus	Beltran-Aguilar et al., 2019
Nitrate reductase	NgNR	*Nannochloropsis gaditana*	Nitrate inducible expression of transgenes 4× higher than WT	eGFP	Jackson et al., 2019
Ammonium transporter	AMT	*Phaeodactylum tricornutum*	Strong inducible expression of transgenes under N starvation	eGFP	Adler-Agnon et al., 2018
Purine permease	PUP	*Phaeodactylum tricornutum*	Strong constitutive expression of transgenes	eGFP	Adler-Agnon et al., 2018
Actin-like 2	Act2	*Phaeodactylum tricornutum*	Weak constitutive and late-inducible expression under N starvation	eGFP	Adler-Agnon et al., 2018
Diacylglycerol acyltransferase	DGAT1	*Phaeodactylum tricornutum*	Weak constitutive and late-inducible expression under N starvation	eGFP	Adler-Agnon et al., 2018
Cyanobacteria arabinose inducible promoter	araBAD	*Escherichia coli*	Strong linear inducible expression of transgenes in response to arabinose	mtGFP	Cao et al., 2017

and promoters, which are inducible in normal situations because they may be used to coordinate transgene verbalization by various activators, hinder adverse effects of fundamental overexpression on the green development's phone absorption and benefit (Zhang et al., 2017). Ribosome-confining districts (RBS) in a similar manner expect a huge part in overseeing conversion created for enhanced quality explanation in fresh and marine water cyanobacterial strains (B. Wang et al., 2018). A list of different promoters used in GM algae development is presented in Table 7.1.

The first results were observed through synthetic algal promoters for high nuclear verbalization in *C. reinhardtii* (Scranton et al., 2016). Subjects in mortified regions of the most prevalent 50 mparting *C. reinhardtii* characteristics have been recognized through POWRS figuring and by discretionarily solidified to 500 nucleotide progressions for making SAPS testing. The SAPS results were implanted on a pBR4 vector spine of a mCherry columnist protein, through hygromycin B quality obliged by B-tubulin sponsor. All in all, 7 out of 25 SAPs drove mCherry enunciation through higher (~2×) than that of the ar1 publicist, showing the capacity of made sponsors for achieving higher recombinant protein yields. Tantamount systems were used for designed sponsor collections for mammalian and bacteriological cell stages, provoking divulgences of key managerial parts and the ability to control recombinant quality verbalization over a wide amazing reach (Brown et al., 2014). Later, the advancement of additional SAP libraries will preferably provoke comparable disclosures and grant more accurate control of planned microalgae characteristics. The extension of an assessment will remain to be focused on the extent of publicists in the innate apparatus compartment for certifying major algal structures.

7.2.3 Agrobacterium-Mediated Change Technique

In the wake of preparation and cloning an ideal verbalization vector, the DNA plasmid vector is implanted with a high-viability algal genome, which may help reduce the time consumed during post-pivotal screening. Various change procedures were fruitful for microalgal changes, including biolistic atom attack, glass globule disrupting, electroporation, silicon carbide stubbles, agrobacterium-mediated, and nanoparticles. The most notable and effective method is electroporation, which works for *C. reinhardtii* nuclear changes where it can yield the transformants a magnitude of 100 times what typical methods provide (Mini et al., 2018). This methodology uses an electrical heartbeat, making cell-divider micropores that license exogenous DNA to enter. Instruments used in this methodology achieve the most limited change efficiencies that have been portrayed (Munoz et al., 2018). Also, they discovered the principle powerful chloroplast change in Nannochloropsis, when electroporation of a nuclear zeroed in on form and was found to be installed in the plastid genome (Gan et al., 2018).

Agrobacterium-mediated change technique has seemed to convey 50 times more productive transformants than the glass-spot strategy techniques for modifying *P. tricornutum*. Despite electroporation remaining the fundamental change method for most microalgal species, various systems provide accommodation in explicit conditions. CRISPR/CRISPR-Associated ribonucleoproteins techniques require vertexing green development cells in a chamber with glass dabs. This method is used to change the fake microRNA and encourage the growth of unsaturated fat substances in *C. reinhardtii* species (Wang et al., 2017). Velmurugan and Deka (2018) observed the changes in haptophytes, particle attack, polyethylene glycol-mediated through the agrobacterium-interceded techniques. It is used to detect very tiny particles when compared to helium particles. CRISPR/CRISPR-Associated ribonucleoproteins are the alternative techniques used to discern the changes in particle attacks in *P. tricornutum*. The observed changes in *P. tricornutum* are twofold, quality and knockouts of a photoreceptor-encoding quality with 65–100% adequacy. Furthermore, it is used to modify the structure of *P. tricornutum* without changing the DNA of the species (Serif et al., 2018). Lately, it was used to convey a human recombinant protein, and the same was used as an anticancer and antiviral expert (El-Ayouty et al., 2019).

7.2.4 Transgene Expression Based on Different Promotors Used in GM Algae

The random changes were happening in the genome while implementing new strategies since it is very important that it may impact the competence of verbalization and the injected area. A couple of systems were made to practice additional site-unequivocal strategies aimed at the growth of the features in the hosts. The changes in *C. reinhardtii* were observed by the usage of homologous recombination through glass-spot and particle-attack techniques. The particle-attack technique gave the best results compared to the glass-spot technique. Keeping the compromise by homologous recombination using round, single-deserted DNA is an alternate method for reducing the subjective fuse of genome DNA. Earlier reports indicated that low transformation speed and transgene verbalization levels are regularly astoundingly poor.

The change of UMP synthase lacking monstrosity via homologous recombination within the unicellular red alga *Cyanidioschyzon merolae* was itemized by Minoda et al. (2004). Similarly, homologous recombination changes were observed in the mixtures nitrate and nitrite reductases of the *Nannochloropsis* species strain. Furthermore, the green *Picoalga ostreococcus tauri* was used in the homologous recombination to yield nitrate reductase and ferritin characteristics. Another model of *Picoalga* species is used for genomic examination and system science due to its incredibly thick haploid genome (Fajardo et al., 2018). The production of extra-chromosomal vectors and episomes is used for transgene overexpression in *P. tricornutum* and *Thalassiosira pseudonana*. Karas et al. (2015) utilized an established strategy for moving the build collected in E. coli diatom cells that are happening at a high change recurrence. The association of yeast-determined arrangement empowered an even replication of episome diatoms (Fajardo et al., 2018).

7.2.5 Changes in GM Algae

The change in genetic ceaselessness and vertical transmission are the essential factors that influence the fruitful usage of the microalgae change. Some microalgae viz.,

Chlamydomonas, *Dunaliella*, *Chlorella*, and *Nannochloropsis*, demonstrate high steadiness after the change. *Thalassiosira weissflogii* is unreliable after a change in the nucleus like that of *Ulva lactuca*, *Porphyra miniata*, *Kappaphycus alvarezii*, and *Gracilaria changii*. The transgenic DNA is very reliable in the *Chlamydomonas*, *Chlorella*, and *Porphyridium* chloroplast, and *Euglena gracilis*. The dependability of transgenic DNA is damaged by species unequivocal since there is a change rate in diverse subnuclear innate frameworks (Fajardo et al., 2018).

7.3 GENOME SEQUENCING IN GM ALGAE

The three important sequenced genomes of microalgal are diatom *Thalassiosira psuedonana*, green alga *Chlamydomonas reinhardtii*, and a second diatom creature *Phaeodactylum tricornutum*. The genomes of three types were helped with appreciation oceanographic, climatologic, and biomedical wonders. The *Micromonas* species have also been inclined to key animals impacted with climate variation (Demory, 2018) and have filled in as models for supplement and viral association of ocean little fish (Bachy, 2018). *Micromonas* and *Ostreococcus* species have extensive plan data from open vaults (Fu et al., 2019).

Presently, the quantity of openly accessible algal sequenced genomes is assessed to be 40–60 (Nelson, 2019). A limited microalgae genealogy is a model living being for varied biological specialties, and many may not sequence. To address this shortcoming, three significant endeavors for arranging countless microalgal genomes were finished. The Marine Microbial Eukaryote Transcriptome Sequencing Project (MMETSP) generated more data in an assorted gathering of uncharacterized animal types (Keeling, 2014). The second task is that ALG-ALL-CODE has sequenced more than 120 genomes from both recently detached and set aside frameworks after the way-of-life assortment focused UTEX and NMCA. This new genome mixture encloses species from different clades, locations, and habitats. The third task is to know about the 10KP genome-sequencing project. The 10KP task will incorporate 10,000 genomes from higher plants, and 3,000 genomes of 10,000 are required from microalgae (Cheng, 2018).

7.3.1 Construction of *Chlamydomonas Reinhardtii* Library

The hereditary screens of C. *reinhardtii* were effectively utilized in the yeast *Saccharomyces cerevisiae* for genomic examinations including drug and target discovery. Before the Chlamydomonas Library Project, there was no freak assortment while using the yeast *Saccharomyces cerevisiae* for genomic examinations. The CLiP project was established by Jonikas, Grossman, Fitz-Gibbon, and Lefebvre. This project includes insertional freak libraries in which the freaks bear an extraordinary, detectable grouping label (Li, 2016). The CLiP freaks were effectively used to find new qualities and delineate novel segments inside the biosynthesis of lipid pathways. The CLiP assortment incorporates qualities associated with abiotic stress resilience, substance affectability, DNA harm fix, and development concentrates under various conditions (Fu et al., 2019).

7.3.2 PCR-Based Amplification Cloning Technique

An ORFeome is an essential asset for a yeast-mixture screen. It can be formulated by PCR magnification cloning techniques. The advancement of ORFeome creation procedures was completed in the previous decade, though it has an expensive and monotonous process for its development (Wiemann, 2016). The advancement in the innovation of blended DNA has the power to authorize business associations to offer supervision to both natural and biotech research communities at lesser costs and higher output through PCR-based amplification cloning techniques. Furthermore, the combined approach allows codon advancement that can be used in settings geared toward the organic entity and its applications. The two cyanobacteria strain *Prochlorococcus marinus* MED4 ORFeomes and NATL1A ORFeomes were combined with twist bioscience and BGI's genome synthesis and editing platform for downstream investigations. The success rate of combinng two species produced 99% more than individual species, whose success rate is only 70% (Ghamsari, 2011; Fu et al., 2019).

7.4 LATEST TECHNOLOGIES FOR GM ALGAE DEVELOPMENT

7.4.1 GT Technology

Directed genome modifications are essential for hereditary examinations and hereditary designing encompassing all parts of science and biotechnological fields. All scientists and biotechnologists were facing difficulties against the overexpression and specific modifications using the random coordination of cloned species. The genome-editing technology (GT) at first progressed in recombinogenic lower eukaryotes through homologous recombination (HR) and had the option to supplant qualities of interest. GT was exhibited effectively in creatures and plants. Be that as it may, GT in these higher living beings has been very difficult, to some extent, since they are not recombinogenic. Recently created strategies, including genome-altering methods, have circumvented this obstacle by designed nucleases, named "atomic scissors," and the subsequent fix of DNA strand breaks creating changes or substitutions of the qualities of premium (Tan et al., 2016).

Zinc-finger nucleases (ZFNs), transcription activator-like effector nucleases (TALENs), and CRISPR / CRISPR-Associated nucleases are formed a single group nuclease. These three nucleases described the alternation of the microalgal genome along with the exiting meganucleases and targetrons in different creatures (Guha et al., 2017). Furthermore, these three nucleases have been authorized to separate the DNA genome and get the quality transformation coming out of it. The genetic design of microalgae is used today for producing biofuels and other biofuel by-products. Genome alteration is the basic concealment of qualities for creating targeted particles and getting quality products. This article depicts the genome alteration of microalgae and its numerous applications in living beings (Jeon et al., 2017). The flow chart for the CRISPR system is shown in Figure 7.2.

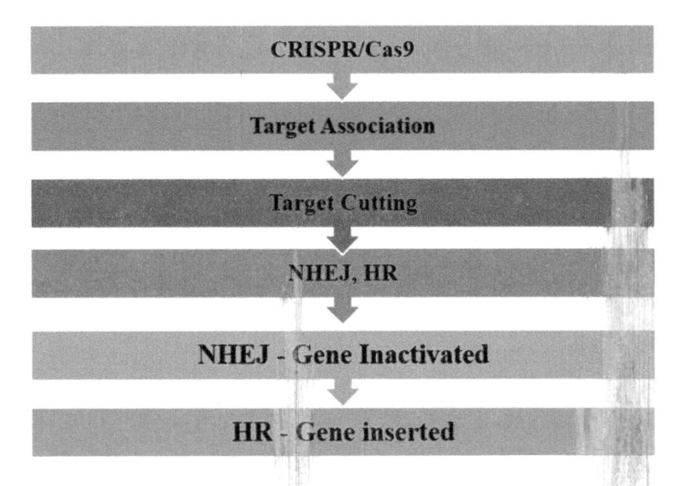

FIGURE 7.2 Flow chart for the CRISPR system.

7.4.2 Gene Regulation

The alga likewise gives an appealing creation host to recombinant proteins, biofuels, and green synthetic compounds (Scranton et al., 2015). Even though *Chlamydomonas* is promptly changeable, transgene articulation from the atomic genome is famously wasteful, and besides, is frequently shaky in that deficiency of transgene articulation happens with time (Yamasaki and Ohama, 2011). Various procedures have been sought to overcome this genuine limit, including the development of mixture advertisers, the incorporation of endogenous introns in the articulation tape, and codon streamlining of the unfamiliar quality (Barahimipour, 2015). Although these procedures mitigated the issue somewhat, the statement of even standard transgenes required exceptional testing. A broader arrangement evaluates the freak strains in terms of improved transgene articulation characteristics. UV mutagenesis screen 15 was helpful for separating UVM4 and UVM11 freak strains used in cell science reads and acting as a tool for undeniable level transgene articulation (Ramos-Martinez et al., 2017). Upgraded transgene articulation connects with significantly expanded record levels, proposing the strains harbor changes in genes, which affect the solid epigenetic transgene quieting in the wild type15. Besides, the transgene hushing pathway explicitly influences exogenously presented DNA. The hushing instrument is also used further to identify how transgene articulation can be improved in headstrong species.

7.4.3 RNA Check

The quality quieting facilitated with RNA or RNA check (RNAi) is a proportioned cycle in a eukaryote of a target assembly through the understanding impediment and RNA degradation (Liang et al., 2018). RNA check pathways stay by and large uncharacterized, even at the level of recognizing urgent quality components in the sequenced genome. An assessment between 14 microalgae species and sequenced genomes indicates that the middle RNAi has all the earmarks of being through and through missing from a couple of green development with minimal nuclear genomes. Similar to *C.* merolae and *G. sulphuraria*, *O. lucimarinus*, *Ostreococcustauri*, *B. prasinos*, and *M. pusilla*, are stable with the hypothesis of eukaryotic evolution (Kim et al., 2018).

The converse hereditary qualities instrument RNAi in microalgae was set up for a couple of animal varieties (Liang et al., 2018). RNAi-induced aggregate and RNA-restricting protein were planned in *C. reinhardtii*, microalgae that reliably set off co-hushing of quality with a selectable RNAi-induced aggregate. Similarly, communicated barrette RNA (hpRNA) comprised the homologous arrangements that were utilized to decrease 40–67% quality in *Dunaliella salina*. The pectin methylesterase (PME) and cellulose synthase (CesA) offer incredible and non-obtrusive methods for examining the explicit proteins in microalgae (Fajardo et al., 2018).

The main measure of work as of late, utilizing RNAi to manage or side road the outflow of qualities, was completed in *C. reinhardtii* (Yamasaki et al., 2013). The measurement recommended base-blending necessities for a little RNA-mediated suppression in microalgae, which are analogous to those of metazoans. An investigation with the truncated light-collecting radio wire quality directed the chlorophyll in *C. reinhardtii*. The RNAi hushing tweaks the statement of light-collecting protein, which gives a robotic comprehension of the chlorophyll receiving wire size guideline by this quality. Besides, prompting RNAi-intervened hushing of the chlorophyllide an oxygenase quality and chlorophyll b levels were somewhat decreased. It was tuning the fringe light-harvesting receiving wires size for the twofold expanded photosynthetic efficiency and a 30% increment in development rate at immersing light forces (Fajardo et al., 2018).

7.4.4 Transcription Factors

Transcriptional designing spotlights the management of numerous qualities in a metabolic pathway simultaneously by designing transcriptional components, for example, transcription factors (TFs); these direct the outflow qualities to explicit themes in the DNA grouping, which makes a connection with RNA polymerase either up-or-down the quality record (Bajhaiya et al., 2017). In contrast to designing explicit qualities, one machine gear-piece in a metabolic pathway, record components can impact various metabolic pathway pieces at the same time. The snapdragon plants gene was embedded with tomato plant gene and observed the degree of anthocyanin and well-being, advancing cell reinforcement. The TFs found in the microalgae lacked explicit capacities detailed for them; be that as it may, proposed transcriptional administrative organizations constructed utilizing RNA-sequenced information have related TFs that control the various parts of cell digestion and their limiting destinations (Romero-Campero et al., 2016).

Fundamental leucine zipper TFs are assumed to be a part of pressure reactions and control the metabolic pathways, like lipid creation (Table 7.2). Lately, *Nannochloropsis* species were developed for strain formation because of their characteristic of producing high lipid levels. TFs in N. salina, which is determined by the endogenous TUB advertiser, and *N. oceanica*, which is determined by the

TABLE 7.2
List of Lipid Biosynthesis Genes

Name	Abbreviation	Origin Organism	Function	Product	References
Basic-region/leucine zipper transcription factor	NobZIP1	*Nannochloropsis oceanica*	Regulates key genes involved in lipid and cell wall carbohydrate metabolism	Lipid biosynthesis	Zhipeng et al., 2018
Basic-region/leucine zipper transcription factor	NsbZIP1	*Nannochloropsis salina*	Involved in regulation of lipid metabolism	Lipid biosynthesis	Kwon et al., 2018
AP2/EREBP transcription factor	Wrinkled1 (Wr1)	*Arabidopsis thaliana*	Regulates lipid accumulation in Arabidopsis seeds	Lipid biosynthesis	Kang et al., 2017
Zn(II)2Cys6 Transcription Factor	ZnCys	*Nannochloropsis gaditana*	Downregulated lipid production conditions	Lipid biosynthesis	Ajjawi et al., 2017

HSP20 advertiser, had been used for building a lipid content in transformant lines of wild sort strains. The type of lipids and classes in N. salina were recognized by using Nile Red fluorescence, chromatography, and HPLC (Kwon et al., 2018). A list of lipid biosynthesis genes is presented in Table 7.2.

As similar to endogenous TFs, the exogenous TFs were used to build lipid in *Nannochloropsis* (Table 7.2). AP2 type TF Wrinkled1 in Arabidopsis (AtWRI1) is used as a regulator for lipid biosynthesis in plants, which is further influenced by the TUB advertiser in *N. salina* (Kang et al., 2017). These outcomes of AtWRI1 TFs could be successful for heterologous articulation in green growth. Currently, CRISPR / CRISPR-Associated was used to find the freak in N. *gaditana* strains, in which TFs were identified by RNA sequenced under nitrogen exhausted conditions (Ajjawi et al., 2017; Sproles et al., 2021).

7.4.5 GM Algae Mutagenesis

Old-style strain-improvement systems depend entirely on mutagenesis that causes hereditary broadening, which is a commonly utilized methodology in microalgae mutagenesis. Irregular mutagenesis expects practically no comprehension of the hidden hereditary qualities or life cycle, and it can be moderately easy to uusee contrasted with carrying out manufactured designing apparatuses or sexual reproduction. Nevertheless, the interaction can be work concentrated since wanted transformations are typically uncommon and aggregate enhancements peripheral, consequently requiring various cycles combined with broad screening. Perhaps the most widely recognized techniques for irregular mutagenesis are the openness to bright radiation; it has promptly accessible, safe, and moderate strategies for producing quality changes. The upgraded transgene articulation has recently demonstrated its value to create *C. reinhardtii* strains UVM4 and UVM11, with no critical positional impacts under the transgene concealment system (Neupert et al., 2020). This methodology was effectively utilized to adjust a business strain of *Desmodesmus armatus* in a wide range of fungicide that decreases the pollution in lakes (Corcoran et al., 2018). In this examination, *D. armatus* existed in different degrees of UV radiation that is further used to expand the portions of fungicide. Enduring confines were consistently evaluated for profitability and chosen less frequently until the maximum encouraging freaks scaled up to 33,000 L raceway lakes and effectively demonstrated compassion toward the fungicide. Notwithstanding UV-illumination, different types of radiation have lately been utilized to cause hereditary varieties. For instance, lipid *Chlorella pyrenoidosa* strain was enhanced by 20% in the wake of presenting cells to 60Co-gamma light (B. Wang et al., 2018).

More than 80% cell mortality from the mutagenesis and 500 freaks were separated with the help of Nile Red fluorescence. In this examination, the last freaks showed powerful development that surpassed the wild-type parent just as contained raised groupings of immersed unsaturated fats. In another examination, a more recent developed strategy for mutagenesis, atmospheric and room-temperature plasma (ARTP), was employed to maximize the formation of hydrogen in *C. reinhardtii* by decreasing the chlorophyll content. ARTP is a novel methodology as it considers various mutagenic specialists to simultaneously follow up on the cells, causing an assortment of mutation types without a moment's delay, significantly expanding the hereditary variety in a populace contrasted with a solitary mutagen. Subsequently, a decrease in chlorophyll content in *C. reinhardtii* was screened by means of recognizing province, which are lighter green, contrasted with the wild sort during the establishment in agar plates (Sproles et al., 2021).

7.4.6 Revamping the Genome

Revamping the genome is an unbelievable approach to the current innate assortment and may be helped with sporadic mutagenesis for strain improvement. Notwithstanding, this approach should be used in strains either through a controllable sexual life cycle or by making protoplasts fit cell mixes. The improved genome is further used in microorganisms and yeasts and could not be implemented in green development. Green development is good for sensual recombination started in research offices; *C. reinhardtii* species is depicted through genome reworking. In one assessment, *C. reinhardtii* strains from freshwater were introduced first to ultraviolet enlightenment and subsequently put through iterative examples of sexual recombination (Takouridis et al., 2015). After three rounds of genome improvement, salt opposition was in the long run extended between 300 and 700 mM NaCl. Fields et al. (2019) assessed the formation of a recombinant protein in *C. reinhardtii* and was extended to 2% of total dissolvable protein in 17 weeks. Further, it was developed by changing a telephone line to convey chloroplast GFP, by then UV-enlightening telephones, and assessing for extending GFP fluorescence by using fluorescence-authorized cell orchestrating (FACS) (Sproles et al., 2021).

7.5 CONCLUSION

The headway of molecular instruments has colossally energized strain planning for biotechnology applications using strains. In the recent past, explicit interests were created in planning carbon progress, which was redirected into biosynthetic pathways like triacylglycerol, polyunsaturated fatty acids, and carotenoids. The planning cells can help their photosynthetic adequacy, even in just growing algal biomass productivity. Planning strategies will presumably be drawn in a zone of investigation for utilizing new resources.

REFERENCES

Adler-Agnon, Z., Leu, S., Zarka, A., Boussiba, S., and Khozin-Goldberg, I. (2018). Novel promoters for constitutive and inducible expression of transgenes in the diatom *Phaeodactylum tricornutum* under varied nitrate availability. *J. Appl. Phycol.*, *30*, 2763–2772.

Ajjawi, I., Verruto, J., Aqui, M., Soriaga, L.B., Coppersmith, J., and Kwok, K. (2017). Lipid production in *Nannochloropsis gaditana* is doubled by decreasing expression of a single transcriptional regulator. *Nat. Biotechnol.*, *35*, 647–652.

Anand, V., Singh, P., Banerjee, C., and Shukla, P. (2017). Proteomic approaches in microalgae: Perspectives and applications. *Biotech.*, *7*, 197.

Bachy, C. (2018). Transcriptional responses of the marine green alga *Micromonas pusilla* and an infecting prasinovirus under different phosphate conditions. *Environ Microbiol.*, *20*, 2898–2912.

Bajhaiya, A.K., Moreira, J.Z., and Pittman, J.K. (2017). Transcriptional engineering of microalgae: Prospects for high-value chemicals. *Trends Biotechnol.*, *35*, 95–99.

Barahimipour, R. (2015). Dissecting the contributions of GC content and codon usage to gene expression in the model alga *Chlamydomonas reinhardtii*. *Plant J.*, *84*, 704–717.

Behera, S., Singh, R., Arora, R., Sharma, N.K., Shukla, M., and Kumar, S. (2015). Scope of algae as third generation biofuels. *Front. Bioeng. Biotechnol.*, *2*, 90. doi:10.3389/fbioe.2014.00090.

Beltran-Aguilar, A.G., Peraza-Echeverria, S., Lopez-Ochoa, L.A., Borges-Argaez, I.C., and Herrera-Valencia, V.A. (2019). A novel salt-inducible CrGPDH3 promoter of the microalga *Chlamydomonas reinhardtii* for transgene overexpression. *Appl. Microbiol. Biotechnol.*, 3487–3499.

Blatti, L., Michaud, J., and Burkart, M. (2013). Engineering fatty acid biosynthesis in microalgae for sustainable biodiesel. *Current Opinion in Chemical Biology*, *17*, 496–505.

Brown, A.J., Sweeney, B., Mainwaring, D.O., and James, D.C. (2014). Synthetic promoters for CHO cell engineering. *Biotechnol. Bioeng.*, *111*, 1638–1647.

Cao, Y.Q., Li, Q., Xia, P.F., Wei, L.J., Guo, N., Li, J.W., and Wang, S.G. (2017). AraBAD based toolkit for gene expression and metabolic robustness improvement in *Synechococcus elongatus*. *Sci. Rep.*, *7*, 2–11.

Cheng, S. (2018). 10KP: A phylodiverse genome sequencing plan. *Gigascience.*, *7*, 1–9.

Chia, S.R., Chew, K.W., Show, P.L., Yap, Y.J., Ong, H.C., Ling, T.C., and Chang, J.S. (2018). Analysis of economic and environmental aspects of microalgae biorefinery for biofuels production: A review. *Biotechnol. J.*, *13*, 1700618. doi:10.1002/biot.201700618.

Corcoran, A.A., Saunders, M.A., Hanley, A.P., Lee, P.A., Lopez, S., Ryan, R., Yohn, C.B. (2018). Iterative screening of an evolutionary engineered Desmodesmus generates robust field strains with pesticide tolerance. *Algal Res.*, *31*, 443–453.

Corteggiani, C.E., Telatin, A., Vitulo, N., Forcato, C.D., Angelo, M., and Schiavon, R. (2014). Chromosome scale genome assembly and transcriptome profiling of *Nannochloropsis gaditana* in nitrogen depletion. *Molecular Plant*, *7*, 323–335.

Crozet, p., Navarro, F.J., Willmund, F., Mehrshahi, P., Bakowski, K., and Lauersen, K.J. (2018). Birth of a photosynthetic chassis: A MoClo toolkit enabling synthetic biology in the microalga *Chlamydomonas reinhardtii*. *ACS Synth. Biol.*, *7*, 2074–2086.

Demory, D. (2018). Picoeukaryotes of the Micromonas genus: Sentinels of a warming ocean. *ISME J.*, *13*, 132–146.

El-Ayouty, Y., El-Manawy, I., Nasih, S., Hamdy, E., and Kebeish, R. (2019). Engineering Chlamydomonas reinhardtii for expression of functionally active human interferonα. *Mol. Biotechnol.*, *61*, 134–144.

Fajardo, Minoda C., Donato, M.D., Carrasco, R., Martınez-Rodrıguez, G., Mancera, J.M., and Fernandez-Acero, F.J. (2018). Advances and challenges in genetic engineering of microalgae. *Reviews in Aquaculture*, 1–17. doi:10.1111/raq.12322.

Fayyaz, M., Chew, K.W., Show, P.L., Ling, T.C., Ng, I., and Chang, J.S. (2020). Genetic engineering of microalgae for enhanced biorefinery capabilities. doi:10.1016/j.biotechadv.2020.107554.

Fields, F.J., Ostrand, J.T., Tran, M., and Mayfield, S.P. (2019). Nuclear genome shuffling significantly increases production of chloroplast-based recombinant protein in *Chlamydomonas reinhardtii*. *Algal Res.*, *41*.

Fu, W., Nelson, D.R., Mystikou, A., Daakour, S., and Salehi-Ashtiani, K. (2019). Advances in microalgal research and engineering development. *Current Opinion in Biotechnology*, *59*, 157–164. doi:10.1016/j.copbio.2019.05.013.

Gan, J., Jiang, X., Han, S., Wang, Y., and Lu. Y. (2018). Engineering the chloroplast genome of oleaginous marine microalga *Nannochloropsis oceanica*. *Front. Plant Sci.*, *9*, 1–6.

Ghamsari, L. (2011). Genome-wide functional annotation and structural verification of metabolic ORFeome of *Chlamydomonas reinhardtii*. *BMC Genomics*, *12*.

Gopal, G.J., and Kumar, A. (2013). Strategies for the production of recombinant protein in *Escherichia coli*. *Protein J.*, *32*, 419–425.

Gould, N., Hendy, O., and Papamichail, D. (2014). Computational tools and algorithms for designing customized synthetic genes. *Front Bioeng. Biotechnol.*, *2*, 41.

Guha, T.K., Wai, A., and Hausner, G. (2017). Programmable genome editing tools and their regulation for efficient genome engineering. *Comput Struct Biotechnol J.*, *15*, 146–160.

Jackson, H.O., Berepiki, A., Baylay, A.J., Terry, M.J., Moore, C.M., and Bibby, T.S. (2019). An inducible expression system in the alga *Nannochloropsis gaditana* controlled by the nitrate reductase promoter. *J. Appl. Phycol.*, *31*, 269–279.

Jeon, S., Lim, J.M., Lee, H.G., Shin, S.E., Kang, N.K., Park, Y., Oh, H.M., Jeong, W.J., Jeong, B., Yong B., and Chang, K. (2017). Current status and perspectives of genome editing technology for microalgae. *Biotechnol Biofuels*, *10*, 267. doi:10.1186/s13068-017-0957-z.

Kang, N.K., Kim, E.K., Kim, Y.U., Lee, B., Jeong, W.J., Jeong, B.R., and Chang, Y.K. (2017). Increased lipid production by heterologous expression of AtWRI1 transcription factor in *Nannochloropsis salina*. *Biotechnol Biofuels*, *10*, 1–14.

Karas, B., Diner, R., Lefebvre, S., McQuaid, J., Phillips, A., and Noddings, C. (2015). Designer diatom episomes delivered by bacterial conjugation. *Nature Communications*, *6*, 6925.

Keeling, P.J. (2014). The Marine Microbial Eukaryote Transcriptome Sequencing Project (MMETSP): Illuminating the functional diversity of eukaryotic life in the oceans through transcriptome sequencing. *PLoS Biol.*, *12*, e1001889.

Kim, J., Liu, L., Hu, Z., and Jin, E. (2018). Identification and functional analysis of the psaD promoter of *Chlorella vulgaris* using heterologous model strains. *Int. J. Mol. Sci.*, *19*(7), 1969.

Koussa, J., Chaiboonchoe, A., and Salehi-Ashtiani, K. (2014). Computational approaches for microalgal biofuel optimization: A review. *Biomed Res Int.*, *649453*. doi:10.1155/2014/649453.

Kwon, S. Kang, N.K., Koh, H.G., Shin, S.E., Lee, B., Jeong, B.R., and Chang, Y.K. (2018). Enhancement of biomass and lipid productivity by overexpression of a bZIP transcription factor in *Nannochloropsis salina*. *Biotechnol. Bioeng.*, *115*, 331–340.

Levering, J. (2016). Genome-scale model reveals metabolic basis of biomass partitioning in a model diatom. *PLoS One*, *11*.

Li, X. (2016). An indexed, mapped mutant library enables reverse genetics studies of biological processes in *Chlamydomonas reinhardtii*. *Plant Cell*, *28*, 367–387.

Liang, M., Zhu, J., and Jiang, J. (2018). High-value bioproducts from microalgae: Strategies and progress. *Critical Reviews in Food Science and Nutrition*, 1455030. doi:10.1080/10408398.2018.

Martella, A., Matjusaitis, M., Auxillos, J., Pollard, S.M., and Cai, Y. (2017). EMMA: An extensible mammalian modular assembly toolkit for the rapid design and production of diverse expression vectors. *ACS Synth. Biol.*, *6*, 1380–1392.

Mini, P., Demurtas, O.C., Valentini, S., Pallara, P., Aprea, G., Ferrante, P., and Giuliano, G. (2018). Agrobacterium-mediated and electroporation-mediated transformation of Chlamydomonas reinhardtii: A comparative study. *BMC Biotechnol.*, *18*, 1–12.

Minoda, A., Sakagami, R., Yagisawa, F., Kuroiwa, T., and Tanaka, K. (2004). Improvement of culture conditions and evidence for nuclear transformation by homologous recombination in a red alga, Cyanidioschyzon merolae 10D. *Plant and Cell Physiology*, *45* (6), 667–671.

Munoz, C.F., Jaeger, L., Sturme, M.H.J., Lip, K.Y.F., and Olijslager, J.W.J. (2018). Improved DNA/protein delivery in microalgae – A simple and reliable method for the prediction of optimal electroporation settings. *Algal Res.*, *33*, 448–455.

Naghshbandi, M.P., Tabatabaei, M., Aghbashlo, M., Aftab, M.N., and Iqbal, I. (2020). Metabolic engineering of microalgae for biofuel production. *Methods in Molecular Biology*, 153–172.

Nelson, D.R. (2017). The genome and phenome of the green alga *Chloroidium* sp. UTEX 3007 reveal adaptive traits for desert acclimatization. *eLife*, *6*. doi:10.7554/eLife.25783.001.

Nelson, D.R. (2019). Potential for heightened sulfur-metabolic capacity in coastal subtropical microalgae. *iScience*, *11*, 450–465.

Ng, I.S., Tan, S.I., Kao, P.H., Chang, Y.K., and Chang, J.S. (2017). Recent developments on genetic engineering of microalgae for biofuels and bio-based chemicals. *Biotechnol. J.*, *12*. doi:10.1002/biot.201600644.

Nymark, M., Sharma, K., Sparstad, T., Bones, M., and Winge, P. (2016). A CRISPR/Cas9 system adapted for gene editing in marine algae. *Scientific Reports*, *6*, 24951.

Poliner, E., Cummings, C., Newton, L., and Farre, E.M. (2019). Identification of circadian rhythms in *Nannochloropsis* species using bioluminescence reporter lines. *Plant J.*, *99*, 112–127.

Ramarajan, M., Fabris, M., Abbriano, R.M., Pernice, M., and Ralph, P.J. (2019). Novel endogenous promoters for genetic engineering of the marine microalga *Nannochloropsis gaditana* CCMP526. *Algal Res.*, *44*,101708.

Ramos-Martinez, E.M., Fimognari, L., and Sakuragi, Y. (2017). High-yield secretion of recombinant proteins from the microalga *Chlamydomonas reinhardtii*. *Plant Biotechnol. J.*, *15*, 1214–1224.

Romero-Campero, F.J., Perez-Hurtado, I., Lucas-Reina, E., Romero, J.M., and Valverde, F. (2016). ChlamyNET: A Chlamydomonas gene co-expression network reveals global properties of the transcriptome and the early setup of key co-expression patterns in the green lineage. *BMC Genomics*, *17*, 1–28.

Scranton, M.A., Ostrand, J.T., Fields, F.J., and Mayfield, S.P. (2015). Chlamydomonas as a model for biofuels and bio-products production. *Plant J.*, *82*, 523–531.

Scranton, M.A., Ostrand, J.T., Georgianna, D.R., Lofgren, S.M., Li, D., and Ellis, R.C. (2016). Synthetic promoters capable of driving robust nuclear gene expression in the green alga *Chlamydomonas reinhardtii*. *Algal Res.*, *15*, 135–142.

Serif, M., Dubois, G., Finoux, A.L., Teste, M.A., Jallet, D., and Daboussi, F. (2018). One-step generation of multiple gene knock-outs in the diatom *Phaeodactylum tricornutum* by DNA-free genome editing. *Nat. Commun.*, *9*, 1–10.

Shin, S., Lim, J., Koh, H., Kim, E., Kang, N., and Jeon, S. (2016). CRISPR/Cas9-induced knockout and knock-in mutations in *Chlamydomonas reinhardtii*. *Scientific Reports*, *6*, 27810.

Sproles, A.E., Fields, F.J., Smalley, T.N., Le, C.H., Badary, A., and Mayfield, S.P. (2021). Recent advancements in the genetic engineering of microalgae. *Algal Research*, *53*, 102158. doi:10.1016/j.algal.2020.102158.

Takouridis, J.S., Tribe, D.E., Gras, S.L., and Martin, G.J.O. (2015). The selective breeding of the freshwater microalga *Chlamydomonas reinhardtii* for growth in salinity. *Bioresour. Technol.*, *184*, 18–22.

Tan, W., Proudfoot, C., Lillico, S.G., and Whitelaw, C.B. (2016). Gene targeting, genome editing: from Dolly to editors. *Transgenic Res.*, *25*(3), 273–287.

Ullmann, J., and Grimm, D. (2019). Algae and their potential for a future bioeconomy, landless food production, and the socio-economic impact of an algae industry. *Org. Agr.* doi:10.1007/s13165-020-00337-9.

Velmurugan, N., and Deka, D. (2018). Transformation techniques for metabolic engineering of diatoms and haptophytes: Current state and prospects. *Appl. Microbiol. Biotechnol., 102*, 4255–4267.

Wang, C., Chen, X., Li, H., Wang, J., and Hu, Z. (2017). Artificial miRNA inhibition of phosphoenolpyruvate carboxylase increases fatty acid production in a green microalga *Chlamydomonas reinhardtii. Biotechnol Biofuels, 10*, 1–11.

Wang, B., Eckert, C., Maness, P.C., and Yu, J. (2018). A genetic toolbox for modulating the expression of heterologous genes in the cyanobacterium *Synechocystis* sp. PCC 6803. *ACS Synth. Biol., 7*, 276–286.

Wang, W., Wei, T., Fan, J., Yi, J., Li, Y., and Wan, M. (2018). Repeated mutagenic effects of 60Co-γ irradiation coupled with high-throughput screening improves lipid accumulation in mutant strains of the microalgae *Chlorella pyrenoidosa* as a feedstock for bioenergy. *Algal Res., 33*, 71–77.

Wiemann, S. (2016). The ORFeome collaboration: A genome-scale human ORF-clone resource. *Nat. Methods, 13*, 191–192.

Weyman, P.D., Beeri, K., Lefebvre, S.C., Rivera, J., McCarthy, J.K., and Heuberger, A.L. (2015). Inactivation of *Phaeodactylum tricornutum* urease gene using transcription activator-like effector nuclease-based targeted mutagenesis. *Plant Biotechnology Journal, 13*, 460–470.

Yamasaki, T., and Ohama, T. (2011). Involvement of Elongin C in the spread of repressive histone modifications. *Plant J., 65*, 51–61.

Yamasaki, T., Voshall, A., Kim, E., Moriyama, E., Cerutti, H., and Ohama, T. (2013). Complementarity to an miRNA seed region is sufficient to induce moderate repression of a target transcript in the unicellular green alga *Chlamydomonas reinhardtii. The Plant Journal, 76*(6), 1045–1056.

Zhang, H., Jing, R., and Mao, X. (2017). Functional characterization of TaSnRK2.8 promoter in response to abiotic stresses by deletion analysis in transgenic Arabidopsis. *Front. Plant Sci., 8*, 1–9.

Zhipeng, L.i., Meng, T., Ling, X., Li, J., Zheng, C., and Shi, Y. (2018). Overexpression of malonyl-CoA: ACP transacylase in *Schizochytrium* sp. to improve polyunsaturated fatty acid production. *J. Agric. Food Chem., 66*, 5382–5391.

8 Immobilized Micro Algae for Removing Wastewater Pollutants and Ecotoxicological View of Adsorbed Nanoparticles – An Overview

D. Suganya
Department of Biology, Gandhigram Rural Institute – Deemed University, Gandhigram, Dindigul Dist, Tamil Nadu, India

M.R. Rajan
Department of Zoology, Micheal job College of Arts and Science for Women, Sulur, Coimbatore Dist, Tamil Nadu, India

Muhilan Mahendhiran
Unidad Multidisciplinaria de Docencia e Investigación, Facultad de Ciencias, Universidad Nacional Autónoma de México (UNAM), Puerto de abrigo s/n, Sisal, Yucatán, México

Kalasalingam School of Agriculture and Horticulture (KSAH), Kalasalingam Academy of Research and Education (Deemed to be University), Krishnankovil, Tamil Nadu, India

Sivakumar Durairaj
Kalasalingam School of Agriculture and Horticulture (KSAH), Kalasalingam Academy of Research and Education (Deemed to be University), Krishnankovil, Tamil Nadu, India

DOI: 10.1201/9781003165101-8

CONTENTS

8.1 INTRODUCTION

The current global situation is suffering through various economic and ecological issues; population increases contribute to depleting the natural environment and producing higher levels of pollution in the last few years. These problems are affecting the environmental biota in different ways. Particularly, water pollution creates a big challenge to solve. In recent years, many new technologies emerged for solving these problems. Phycoremediation is a one of the techniques for purifying the wastewater (Batan et al., 2010 and Panga Kiran Kumar et al., 2018).

Phycoremediation is a biotechnology due to its vast environmental benefits; it emerged over the past few decades. There has a gentle rise worldwide within the number of phycoremediation algae. It involves the remediation of contaminants during the aquatic body using micro and macro algae. Generally, algae fix carbon dioxide by photosynthesis and take away excess nutrients effectively and economically. It eliminates all pathogens and xenobiotic materials from wastewater. Phycoremediation is used for improving water quality, and additionally, photosynthetically produced oxygen can relive biological oxygen demand (BOD) within the wastewater (Volesky, 1999). The high risk of accidental release of metal and nonmetal pollutants into the environment can cause health, safety, and ecological problems. Solutions in these environmental concerns include employing algae for remediation, particularly for the removal of wastes as nutrients, which is the basis for algal blooming and enzymatically degrading the water pollutants (Sivasubramanian et al., 2009). This type of algal metabolism can utilize the nitrogen and phosphorous compounds. Chlorella, scenedesmus, synechoccystis, gloeocapsa, chroococcus, anabaena, lyngbya, oscillatoria, spirulina, etc. are generally used for wastewater treatment in most rivers and lakes because of organic and industrial wastes (Ezenweani Raymond et al., 2021).

Massive industrialization and development may present a challenge for the safe disposal of industrial effluents in their existing form; for instance, liquid or solid wastes. These waste material compounds damage aquatic life; in particular, overloaded metals can kill microorganisms (Parameswari et al., 2010). So, prior to discharging industrial wastewater into waste bodies, removing the most nutritious compounds (nitrogen and phosphorous) is usually obligatory, even though it is not performed in many cases, especially in all developing countries (Dueñas et al., 2003). For this wastewater treatment, the phycoremediation process involves nanotechnological techniques. Among biological methods proposed for removing the wastewater pollutants, immobilized microalgae seems to be advantageous.

A worldwide trend of immobilizing microbes in an assortment of matrices is employed for a good sort of biotechnological applications that started over 40 years ago (Lebeau and Robert, 2006; Moreno-Garrido, 2008). Immobilization is the physical or chemical materials to limit cells or enzymes within the confined space area, keeping them active and repeatedly used. Due to the microbial density, fast response, low sludge production, resistance to environmental impact, and easy-to-control reaction process, immobilized microbes have been widely researched and applied in wastewater treatment (Xia Bing et al., 2010). Recently, some magnetic nanoparticles also involved as an immobilizing agent can accumulate in the water bodies and microalgae. These agents are transferred to the food chain from microalgae to higher vertebrates. Many researchers are finding the ecotoxicological view of nanoparticle toxicity on invertebrates and vertebrate animals. This book chapter provides an overview of different magnetic nanoparticles for the removal of organic and inorganic matters from the wastewater and its toxicological effects on human beings and other biotic environment.

8.2 PHYCOREMEDIATION OF GREEN SYNTHESIZED NANOPARTICLES FOR WASTEWATER TREATMENT

In recent days, large numbers of physical, chemical, and biological or green nanotechnogical methods are available for the synthesis of metal and nonmetal oxide nanoparticals. Among them, the biological method of green synthesis has attracted considerable attention with the aid of novel eco and convenient biological materials, namely fungi, bacteria, algae (Figure 8.1), biomolecules, and plants (Suganya et al., 2018). According to some recent studies, freshwater algae like *Scenedesmus quadricauda, Chlorella vulgaris, Selenastrum capricornutum*, and *Scenedesmus platydiscus Scenedesmus quadricauda, Chlorella vulgaris, Selenastrum* are capable of accumulating and degrading polycyclic aromatic hydrocarbons. Algae systems have conventionally been employed as a tertiary wastewater treatment process. These systems have

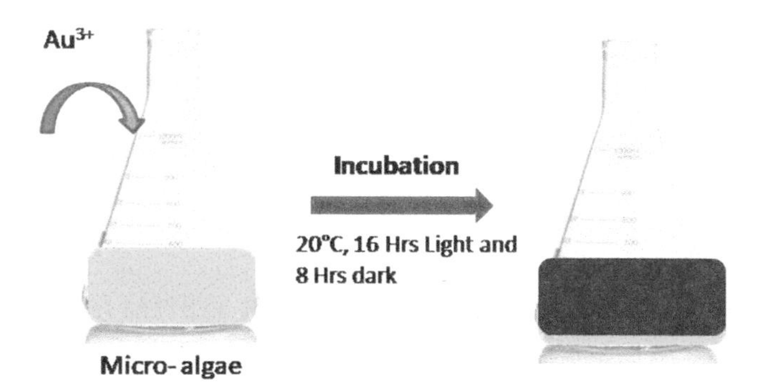

FIGURE 8.1 Green synthesis of metal nanoparticles.

FIGURE 8.2 Common application of metal-oxide nanoparticles.

also been proposed as a possible secondary treatment system. Algae are often utilized in treatment processes for various purposes, a number of which are used for the removal of coliform bacteria, reduction in both chemical and biochemical oxygen demand, and the removal of heavy metals (Rao et al., 2011). Nano-sized absorbents such as metal and metal-oxide nanoparticles, graphene and carbon tubes, and some nanofibers are widely used for improving treatment of water and wastewater. These nanomaterials are higher-performance adsorbents; in particular, zinc oxide (Chouchene et al., 2017) and titanium oxide nanoparticles (Syngouna et al., 2017) have exhibited favorable sorption toward organic and inorganic pollutants. (Figure 8.2).

The major classes of wastewater pollutants are heavy metal ions and some dyes. Water containing such pollutants should not to be used for drinking purposes without purification. Once these types of heavy metal ions and nanoparticles enter water, it is extremely difficult to completely treat or purify the water (Rai et al., 2005). These aquatic pollutants are hazardous for all living organisms and strongly affect the ecosystem. Therefore, these hazardous pollutants must be eliminated from contaminating water to stop their harmful effects on humans and the environment. Water systems face many diverse challenges today. Nearly 800 million people worldwide lack access to clean drinking water (Herschy, 2012).

8.3 ROLE OF NANOPARTICLES IN WATER SYSTEM

Nanomaterials are expected to be present in the sedimentation stage in water environments with very low concentrations from mg/l to μg/L according to previous studies. Some researchers have reported that nanoparticles can deposit in water systems and impact the gene-level toxicity tested on zebrafish (*Danio rerio*) (Suganya et al., 2018). These nanoparticles are aggregate and sediment in the bottom of water bodies; they may occur naturally in the presence of suspended or dissolved substances in water, particularly nature organic matters (Weinberg et al., 2011). Meanwhile, nanoparticles pH, stability, dissolved rate, electrolyte species, and concentration can play a vital role. Other characteristics of water could make aggregation of nanoparticles by changing the neutralization, bridging, and electrical double layer compression; other mechanisms also cause nanoparticles to be more stable. Aggregated nanoparticles may enter the food chain and transfer from lower organisms to higher vertebrates (Hyung et al., 2006; Zhang et al., 2009; Keller, 2010). Nanoparticles may directly enter the aquatic system through industrial effluents or wastewaters, directly or indirectly mixing with surface runoff from soils. These dissolved or mixed nanoparticles may release highly toxic compounds into the aquatic environment. Meanwhile, these hazardous nanoparticles can conglomerate with coexisting or homoaggregate or combine with other organic compounds or minerals to significantly alter the interactions with biota and potential toxic in the environment (Chen et al., 2011; Jang et al., 2014; Rocha et al., 2015).

8.4 ENVIRONMENTAL IMPACT OF NANOPARTICLES

Nowadays silver nanoparticles are widely used in the all types of consumer products, such as water treatment due to their antimicrobial properties. All types of cosmetics are used increasingly, raising ecological concerns because of the release of silver nanoparticles into the aquatic environment. Once zero-valent silver is released into the aquatic system, it may be oxidized to $Ag+$ and the cation liberate or persist as silver nanoparticles. These types of aggregation appear to influence the initiation of other processes, including precipitation, adsorption, and dissolution. These aggregated particles may transfer to the aquatic system. Many researchers have reported that the aggregation is confined or very limited and the particles in the system may stay suspended (Liu et al., 2014).

On other hand, aquatic nanoparticles include adsorption/desorption, sedimentation, degradation, and reactions provide more stability to nanoparticles. Sometimes nanoparticles maybe bind with toxic substances during their exposure in the aquatic conditions. They will enter aquatic animals and accumulate in the animal bodies and then transfer to human bodies before their removal. This contamination is a major reason to begin considering how to remove nanoparticles from aquatic bodies, especially from wastewater and treated wastewater (Figure 8.3).

8.4.1 Nanotoxicity on Fish Species

Nanotoxicology is the branch of nanoscience that deals with the toxicity caused by nanoparticles. It is difficult to test and estimate the negative effects of nanomaterials on living organisms (Ali et al., 2011). Nanotoxicity of these particles is broadly classified into biological toxicity and environmental toxicity. In case of biological toxicity, nanoparticles enter organisms via dermal, subcutaneous, oral, intravenous, intraperitoneal, and inhalation routes (Fu et al., 2014; Malysheva et al., 2015). In

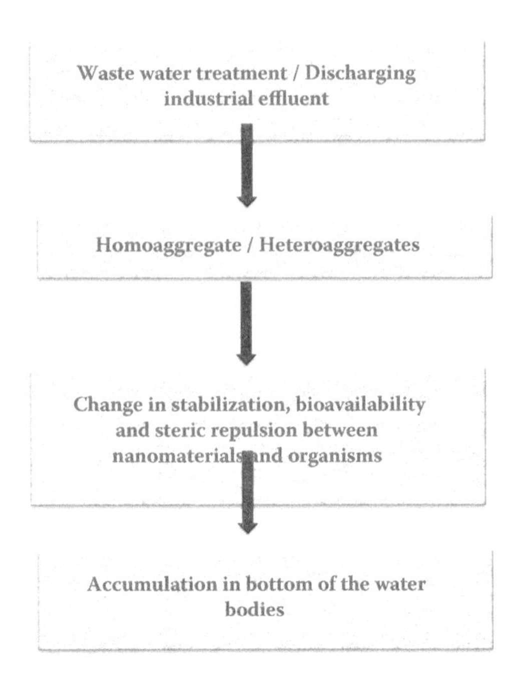

FIGURE 8.3 Nanoparticles pathways in the water system.

environmental toxicity, inorganic and organic nondegradable nanoparticles deposit in the ground and groundwater and lead to water pollution (Hussain et al., 2015; Pour Gashtasbi, 2015). Ecosystems are polluted in many ways; in particular, the aquatic ecosystem is extremely polluted by nanotechnological-related industrial wastes, drainages, garbage waste, etc. (Jahan et al., 2017).

Currently, nanomaterials do not have any permissible limits to be discharged into the environment, and the dissolution of nanoparticles may release potentially toxic compentents into the environment. Moreover, the route of entry is different from environment to environment; if not systematically monitored, these routes may cause a negative effect on living organisms. In the environment, nanoparticles can undergo a number of potential transformations that depend on the properties of both nanoparticles and the receiving medium. Nanoparticles can be added to soils directly through fertilizers or plant-protection products, indirectly through industrial discharge or the disposal of wastewater treatment effluents, or indirectly through surface runoff from soils. The metal nanoparticles are also discharged into the water bodies through the waste materials released from fabric and cosmetic industries (Gottschalk and Nowack, 2011). These nanopollutants cause toxic effects to aquatic organisms, such as phytoplanktons and zooplanktons, through the primary food chain (Karthikeyeni et al., 2015). Likewise in recent years, many metallic nanopolutants like iron, zinc, aluminium, and nickel, which are used in the removal of dyes and toxic materials from wastewater, also affect the aquatic ecosystem. Among the metal-oxide nanoparticles, iron-oxide nanoparticles are used to remove heavy metals (lead, chromium, arsenic, copper, etc.) from wastewater. Iron-oxide nanoparticles are used for the removal or degradation of a wide range of chemical pollutants, such as chlorinated solvents, chlorinated pesticides, ß-lactam, and nitroimidazole-based antibiotics, azo dyes, polychlorinated biphenyls, inorganic anions, including nitrate and perchlorate, nitroamines, nitroaromatics, alkaline earth metals, barium and beryllium, p-chlorophenol, organophosphates, transition metals including chromium, copper, cobalt, molybdenum, nickel, lead, silver, vanadium and technetium, post-transition metals, including zinc and cadmium, metalloids including arsenic, and actinides including uranium and plutonium (Crane and Scott, 2012).

8.5 REMOVAL OF METAL PARTICLES WITH MICROALGAE IN THE WATER SYSTEM

Industrial wastewater is rich in heavy metals like copper, gold, ferrous, lead and etc. These heavy metals, present in mixture forms of liquid sedimentation, are slowly released into the aquatic environment, making it a constant pollution for aquatic animals. Those pollutants produce a heavy loss in animal and plant numbers. The common ways of removing pollutants are metal recovery, absorptions, and removal using polymers; some precipitation methods also are available. Table 8.1 shows the list of heavy metal pollutant removal by microalgae. *Chlorella vulgaris* cells are directly released for removal of copper from the wastewater. Higher density, specific amount of uptake characters are a reason to select the *Chlorella vulgaris*. Some 90% of gold particles immobilized by microalgae (alginate immobilized *Chlorella homosphaera*) were absorbed and 40% absorbed by alginate. In this way, recovery from any environment can be a profitable venture, as reported by da Costa and Leite (1991). Abdel Hameed (2006) reported that the *Chlorella vulgaris* and alginate matrix were more effective for the removal of lead (>90%) and recovery (100%) yield achieved. Alginate matrix adsorption plays a minor role contributed by the microalgae.

TABLE 8.1
Removal of Heavy Metals by Immobilized Microalgae

S. No.	Pollutant	Immobilizing Material	Microalgae Species	References
1	Copper	Alginate	*Anabaena doliolum; Chlorella vulgaris; C. miniata; Nannochloropsis gaditana; Scenedesmus quadricauda; Tetraselmis chui*	Bayramoğlu and Arica (2009); Lau et al. (1998), Mehta and Gaur (2001); Moreno-Garrido et al. (2002, 2005); Rai and Mallick (1992); Tam et al. (2009), Chu and Hashim (2001)
2	Gold	Alginate	*Chlorella homosphaera*	da Costa and Leite (1991)
3	Ferous	Alginate	*Anabaena doliolum; Chlorella vulgaris*	Rai and Mallick (1992)

(Continued)

TABLE 8.1 (Continued)
Removal of Heavy Metals by Immobilized Microalgae

S. No.	Pollutant	Immobilizing Material	Microalgae Species	References
4	Lead	Alginate, Luffa cylindrica sponge	*Chlorella vulgaris; Chlamydomonas reinhardtii Luffa cylindrica sponge Chlorella sorokiniana*	Abdel Hameed (2006); Bayramoğlu et al. (2006)
5	Manganese	Alginate	*Chlorella salina*	Granham et al. (1992)
6	Mercury	Silica gel, Alginate & Agarose	*Chlorella vulgaris, Chlorella emersonii; Chlamydomonas reinhardtii, Chlorella emersonii*	Bayramoğlu et al. (2006); Wilkinson et al. (1990)
7	Zinc	Alginate	*Chlorella homosphaera; C. miniata; C. salina; Nannochloropsis gaditana; Scenedesmus quadricauda*	Bayramoğlu and Arica (2009); da Costa and Leite (1991); Granham et al. (1992); Moreno-Garrido et al. (2002); Tam et al. (2009)
8	Uranium	Polyacrylamide	*Chlorella regularis*	Nakajima et al. (1982)
9	Mixture of Cu, Fe, Ni, Zn	Polysulphone and epoxy resin	*Phormidium laminosum*	Blanco et al. (1999)
10	Cadmium	Alginate	*Chlorella homosphaera; C. vulgaris; Chlamydomonas reinhardtii; Oscillatoria sp.; Tetraselmis chuii*	Bayramoğlu et al. (2006); Cañizares-Villanueva et al. (2000); da Costa and Leite (1991); Katircioğlu et al. (2008); Moreno-Garrido et al. (2005)

Many studies reported the removal of heavy metals by using of immobilized microalgea, but the toxicological view of studies is very limited. Particularly, nanoscale-level aquatic sedimentation, bioaccumulation, and its toxicological information on aquatic biota is very limited.

8.6 BIOACCUMULATION OF NANOPARTICLES ON FISH SPECIES

Research on fish revealed that the nanoparticles are toxic in both high and low concentrations (Handy et al., 2011; Shaw and Handy 2011). Most of the studies concerning iron-oxide nanoparticles (Fe_3O_4) are toxic because of their attractive medicinal properties (Soenen and De Cuyper, 2010; Mahmoudi et al., 2011; Diana et al., 2013; Ebrahiminezhad et al., 2015; Shen et al., 2015) and dose-dependent toxic effects were observed on aquatic organisms, particularly on fish species. Furthermore, nanoparticle toxicity on fish may easily affect their gill surface because the gill is the primary organ making direct contact with the aquatic environment. Possible discharge of wastewater containing the heavy metal nanoparticles from the nanotechnological industry poses risk to the aquatic environment (Wang et al., 2015), and it automatically deposits first in fish gills and transfers to other organs via the circulatory system. Bioacculation of nanoparticle is the deposition of particles in fish organs through all possible route exposures (Dalai et al., 2014). Nanoparticles accumulation can lead to a distribution of ionic pathway and oxidative stress. It leads to call injury in fish organs (Lushchak, 2011). The distribution of ionic pathways automatically changed the level of hematological compents. Hematological parameters are widely used as pathophysiological indicators to diagnose the morphological and functional alternations in fish exposed to various nanopollutants (Adhikari et al., 2004).

8.7 CONCLUSION

In this 21st century, green nanotechnology will have a crucial role in nanomedicine and other biotechnological applications. The integration of nanotechnology into biotechnology has created a replacement science nanotechnology. The era of nanotechnology may be a prominent gift for science worldwide. This new technology corresponds to the present scientific urge for improving the prevailing strategies for nanoparticle biosynthesis and inventing new ones. Biogenic synthesis of nanoparticles can reduce environmental contaminations and reduce human health hazards resulting from currently used conventional manufacturing processes. Immobilized microalgea are often utilized in wastewater treatment for a variety of purposes, including reduction of biological oxygen demand (BOD), removal of nitrogen and /or phosphorous, inhibition of coliforms, and removal of heavy metals from wastewater. Hence, microalgal-mediated nanoparticle wastewater treatment, through biological and physicochemical mechanisms, could represent biological treatment to purify wastewater. It is already being demonstrated that nanoparticles have the potential to accumulate intracellular concentrations. Currently, research goes on to supply recombinant proteins from algae. More research is required not only on nanoparticles but also on raised understanding of the mechanism of nanoparticles formation in immobilized stage by microalgae. Because green technology may less toxic when compared to chemical-synthesized nanoparticles, various aspects of the biotechnological potential of green-synthesized nanoparticles are yet to be clarified.

REFERENCES

Abdel Hameed, M.S. (2006). Continuous removal and recovery of lead by alginate beads, free and alginate-immobilized Chlorella vulgaris. *Afr. J. Biotechnol.*, *5*, 1819–1823.

Adhikari, S., Sarkar, B., Chatterjee, A., Mahapatra, C.T., and Ayyappan, S. (2004). Effects of cypermethrin and carbofuran on certain hematological parameters and prediction of their recovery in a freshwater teleost, *Labeo rohita* (Hamilton). *Ecotoxicology and Environmental Safety*, *58*, 220–226.

Ali, J., Biazar, E., Jafarpour, M., Montazeri, M., Majdi, A., Aminifard, S., Zafari, M., Hanie, R.A., and Rad, J.G.H. (2011). Nanotoxicology and nanoparticle safety in biomedical designs. *International Journal of Nanomedicine*, *6*, 1117–1127.

Batan, L., Jason, Q., Bryan, W., and Thomas B. (2010). Net energy and greenhouse gas emission evaluation of biodiesel derived from microalgae Environ. *Sci. Techno.*, *44*(20), 7975–7980.

Bayramoğlu, G., and Arica, M.Y. (2009). Construction a hybrid biosorbent using Scenedesmus quadricauda and Ca-alginate for biosorption of Cu(II), Zn(II) and Ni(II): Kinetics and equilibrium studies. *Bioresour. Technol.*, *100*, 186–193.

Bayramoğlu, G., Tuzun, I., Celik, G., Yilmaz, M., and Arica, M.Y. (2006). Biosorption of mercury (II), cadmium(II) and lead(II) ions from aqueous system by microalgae Chlamydomonas reinhardtii immobilized in alginate beads. *Int. J. Miner. Process*, *81*, 35–43.

Bing, X., Quan-sheng, Z., and Yang, Q. (2010). The research of different immobilized microorganisms technologies and carriers in sewage disposal. *Science & Technology Information*, *1*(2020), 698–699.

Blanco, A., Sanz, B., Llama, M.J., and Serra, J.L. (1999). Biosorption of heavy metals to immobilized Phormidium laminosum biomass. *J. Biotechnol.*, *69*, 227–240.

Cañizares-Villanueva, R.O., Martínez-Jerónimo, F., and Espinosa-Chávez, F. (2000). Acute toxicity to Daphnia magna of effluents containing Cd, Zn, and a mixture Cd–Zn, after metal removal by Chlorella vulgaris. *Environ. Toxicol.*, *15*, 160–164.

Chen, J., Xiu, Z., Lowry, G.V., and Alvarez, P.J. (2011). Effect of natural organic matter on toxicity and reactivity of nano-scale zero-valent iron. *Water Res.*, *45*(5), 1995–2001.

Chouchene, B., Chaabane, T.B., Mozet, K., Girot, E., Corbel, S., Balan, L., Medjandi, G., and Schneider, R. (2017). Porous Al-doped ZnO rods with selective adsorption properties. *Appl Surf Sci.*, *409*, 102–110.

Chu, K.H., and Hashim, M.A. (2001). Desorption of copper from polyvinyl alcoholimmobilized seaweed biomass. *Acta Biotechnologica*, *21*, 295–306.

Crane, R.A., and Scott, T.B. (2012). Nanoscale zero-valent iron: Future prospects for an emerging water treatment technology. *Journal of Hazardous Materials*, *211–212*, 112–125.

da Costa, A.C.A., and Leite, S.G.F. (1991). Metals biosorption by sodium alginate immobilized Chlorella homosphaera cells. *Biotechnol. Lett.*, *13*, 559–562.

Dalai, S., Iswarya, V., Bhuvaneshwari, M., Pakrashi, S., Chandrasekaran, N., and Mukherjee, A. (2014). Different modes of TiO_2 uptake by *Ceriodaphnia dubia*: Relevance to toxicity and bioaccumulation. *Aquatic Toxicology*, *152*, 139–146.

Diana, V., Bossolasco, P., Moscatelli, D., Silani, V., and Cova, L. (2013). Dose dependent side effect of superparamagnetic iron oxide nanoparticle labeling on cell motility in two fetal stem cell populations. *PLOS One*, *8*(11), 1–12.

Dueñas, J.F., Alonso, J.R., Rey, A.F., and Ferrer, A.S. (2003). Characterisation of phosphorous forms in wastewater treatment plants. *J. Hazard. Mater.*, *97*, 1–3.

Ebrahiminezhad, A., Rasoul-Amini, S., Kouhpayeh, A., Davaran, S., Barar, J., and Ghasemi, Y. (2015). Impacts of amine functionalized iron oxide nanoparticles on HepG2 cell line. *Current Nanoscience*, *11*(1), 113–119.

Fu, P.P., Xia, Q., Hwang, H.M., Ray, P.C., and Yu, H. (2014) Mechanisms of nanotoxicity: Generation of reactive oxygen species. *Journal of Food and Drug Analysis*, *22*(1), 64–75.

Gottschalk, F., and Nowack, B. (2011). The release of engineered nanomaterials to the environment. *Journal of Environmental Monitoring*, *13*(5), 1145–1155.

Granham, G.W., Codd, G.A., and Gadd, G.M. (1992). Accumulation of cobalt, zinc and manganese by the esturine green microalga Chlorella salina immobilized in alginate microbeads. *Environ. Sci. Technol.*, *26*, 1764–1770.

Handy, R.D., Al-Bairuty, G., Al-Jubory, A., Ramsden, C.S., Boyle, D., Shaw, B.J., and Henry, T.B. (2011). Effects of manufactured nanomaterials on fishes: A target organ and body systems physiology approach. *Journal of Fish Biology*, *79*(4): 821–853.

Herschy, R.W. (2012). Water quality for drinking: WHO guidelines Encycl. *Earth Sci. Ser.*, 876–883. doi:10.1007/978-1-4020-4410-6_184.

Hussain, S.M., Warheit, D.B., Ng, S.P., Comfort, K.K., Grabinski, C.M. and Braydich-Stolle, L.K. (2015). At the crossroads of nanotoxicology *in vitro*: Past achievements and current challenges. *Toxicological Science*, *147*(1), 5–16.

Hyung, H., Fortner, J.D., Hughes, J.B., and Kim, J. (2006). Natural organic matter stabilizes carbon nanotubes in the aqueous phase, *Environ. Sd. Technol.*, *41*(2006), 179–184.

Jahan, S., Alias, Y.B., and AbuBakar, A.F.B. (2017). Reviews of the toxicity behavior of five potential engineered nanomaterials (ENMs) into the aquatic ecosystem. *Toxicology Reports*, *4*, 211–220.

Jang, M.H., Bae, S.J., Lee, S.K., Lee, Y.J., and Hwang, Y.S. (2014). Effect of material properties on stability of silver nanoparticles in water. *J Nanosci Nanotechnol.*, (12), 9665–9669.

Karthikeyeni, S., Siva Vijayakumar, T., Vasanth, S., Arul Ganesh, S., Vignesh, V., Akalya, J., Thirumurugan, R., and Subramanian, P. (2015). Decolourisation of direct orange dye by ultra sonication using iron oxide nanoparticles. *Journal of Experimental Nanoscience*, *10*(3), 199–208.

Katircioğlu, H., Aslim, B., Türker, A.R., Atici, T., and Beyatli, Y. (2008). Removal of cadmium(II) ion from aqueous system by dry biomass, immobilized live and heat-inactivated Oscillatoria sp. H1 isolated from freshwater (Mogan Lake). *Bioresour. Technol.*, *99*, 4185–4191.

Keller, A.A. (2010). Stability and aggregation of metal oxide nanoparticles in natural aqueous matrices, *Environ. Sei. Technol.*, *44*(2010), 1962–1967.

Lau, A., Wong, Y.S., Zhang, T., and Tam, N.F.Y. (1998). Metal removal studied by a laboratory scale immobilized microalgal reactor. *J. Environ. Sci. China*, *10*, 474–478.

Lebeau, T., and Robert, J.M. (2006). Biotechnology of immobilized micro-algae: A culture technique for the future. In Rao, S. (Ed.), *Algal cultures, analogues of blooms and applications* (pp. 801–837). Science Publishers, Enfield, NH.

Liu, Y., Tourbin, M., Lachaize, S., and Guiraud, P. (2014). Nanoparticles in wastewaters: Hazards, fate and remediation. *Powder Technology*, *255*, 156–249. ISSN 0032-5910.

Lushchak, V.I. (2011). Adaptive response to oxidative stress: Bacteria, fungi, plants and animals. *Comparative Biochemistry and Physiology*, *153* (2), 175–190.

Mahmoudi, M., Laurent, S., Shokrgozar, M.A., and Hosseinkhani, M. (2011). Toxicity evaluations of superparamagnetic iron oxide nanoparticles: Cell "vision" versus physicochemical properties of nanoparticles. *ACS Nano*, *5*(9), 7263–7276.

Malysheva, A., Lombi, E., and Voelcker, N.H. (2015). Bridging the divide between human and environmental nanotoxicology. *Nature Nanotechnology*, *10*, 835–844.

Mehta, S.K., and Gaur, J.P. (2001). Removal of Ni and Cu from single and binary metal solutions by free and immobilized Chlorella vulgaris. *Eur. J. Protistol.*, *37*, 261–271.

Moreno-Garrido, I. (2008). Microalgae immobilization: Current techniques and uses. *Bioresour. Technol.*, *99*, 3949–3964.

Moreno-Garrido, I., Codd, G.A., Gadd, G.M., and Lubián, L.M. (2002). Cu and Zn accumulation by calcium alginate immobilized marine microalgal cells of Nannochloropsis gaditana (eustigmatophyceae). *Cienc. Mar.*, *28*, 107–111.

Moreno-Garrido, I., Campana, O., Lubián, L.M., and Blasco, J. (2005). Calcium alginate immobilized marine microalgae: Experiments on growth and short-term heavy metal accumulation. *Mar. Pollut. Bull.*, *51*, 823–929.

Nakajima, A., Horikoshi, T., and Sakaguchi, T. (1982). Recovery of uranium by immobilized microorganisms. *Appl. Microbiol. Biotechnol.*, *16*, 88–91.

Panga Kiran Kumar, S., Krishna, V., Kavita Verma, K., Pooja, D., Bhagawan, V., and Himabindu (2018). Phycoremediation of sewage wastewater and industrial flue gases for biomass generation from microalgae. *South African Journal of Chemical Engineering*, *25*, 133–146.

Parameswari, E., Lakshmanan, A., and Thilagavathi, T. (2010). Phycoremediation of heavy metals in polluted water bodies. *Electronic Journal of Environmental, Agricultural and Food Chemistry*, *9*(4), 808–814.

Pour Gashtasbi, G. (2015). Nanotoxicology and challenges of translation. *Nanomedicine (Lond.)*, *10*, 3121–3129.

Rai, H.S., Bhattacharyya, M.S., Singh, J., Bansal, T.K., Vats, P., and Banerjee, U.C. (2005). Removal of dyes from the effluent of textile and dyestuff manufacturing industry: A review of emerging techniques with reference to biological treatment. *Crit. Rev. Environ. Sci. Technol.*, *35*(3), 219–238. doi:10.1080/10643380590917932.

Rai, L.C., and Mallick, N. (1992). Removal and assessment of toxicity of Cu and Fe to Anabaena doliolum and Chlorella vulgaris using free and immobilized cells. *World J. Microb. Biotechnol.*, *8*, 110–114.

Rao, P.H., Kumar, R.R., Raghavan, B.G., Subramanian, V.V., and Sivasubramanian, V. (2011). Application of phycoremediation technology in the treatment of wastewater from a leatherprocessing chemical manufacturing facility. *Water SA*, *37*(1), 7–14.

Rocha, T.L., Gomes, T., Sousa, V.S., Mestre, N.C., and Bebianno, M.J. (2015). cotoxicological impact of engineered nanomaterials in bivalve molluscs: An overview. *Mar. Environ. Res.*, *111*, 74–88.

Shaw, B.J., and Handy, R.D. (2011). Physiological effects of nanoparticles on fish: A comparison of nanometals versus metal ions. *Environment International*, *37*(6), 1083–1097.

Shen, Y., Huang, Z., Liu, X., Qian, J., Xu, J., Yang, X., Sun, A., and Ge, J. (2015). Iron-induced myocardial injury: An alarming side effect of super paramagnetic iron oxide nanoparticles. *Journal of Cellular and Molecular Medicine*, *19*(8), 2032–2035.

Sivasubramanian, V., Subramanian, V.V., Raghavan, B.G., and Ranjithkumar, R. (2009). Large scale phycoremediation of acidic effluent from an alginate industry. *Science Asia*, *35*, 220–226.

Soenen, S.J., and De Cuyper, M. (2010). Assessing iron oxide nanoparticle toxicity in vitro: Current status and future prospects. *Nanomedicine*, *5*(8), 1261–1275.

Suganya, D., Ramakritinan, C.M., and Rajan, M.R. (2018). Adverse effects of genotoxicity, bioaccumulation and ionoregulatory modulation of two differently synthesized iron oxide nanoparticles on Zebrafish (*Danio rerio*). *J Inorg. Organomet. Polym.*, *28*, 2603–2611. doi:10.1007/s10904-018-0935-3.

Syngouna, V.I., Chrysikopoulos, C.V., Kokkinos, P., Tselepi, M.A. and Vantarakis, A. (2017). Cotransport of human adenoviruses with clay colloids and TiO_2 nanoparticles in saturated porous media: Effect of flow velocity. *Sci. Total Environ.*, *598*, 160–167.

Tam, N.F.Y., Wong, Y.S., and Wong, M.H. (2009). Novel technology in pollutant removal at source and bioremediation. *Ocean Coast. Manage.*, *7*, 368–373.

Volesky, B. (1999). Biosorption for the next century. In Ballester, A., and Amils, R. (Eds.), *Biohydrometallurgy and the environment toward the mining of the 21st century. International Biohydrometallurgy Symposium Proceedings* (pp. 161–170), vol. B. Elsevier Sciences, Amsterdam, Netherlands.

Wang, Z., Fang, C., and Mallavarapu, M. (2015). Characterization of iron–polyphenol complex nanoparticles synthesized by sage (*Salvia officinalis*) leaves. *Environmental Technology and Innovation*, *4*, 92–97.

Weinberg, H., Galyean, A., and Leopold, M. (2011). Evaluating engineered nanoparticles in natural waters, *TrAC Trends Anal. Chem.*, *30*, 72–83.

Wilkinson, S.C., Goulding, K.H., and Robinson, P.K. (1990). Mercury removal by immobilized algae in batch culture systems. *J. Appl. Phycol.*, *2*, 223–230.

Zhang, Y., Chen, Y., Westerhoff, P., and Crittenden, J. (2009). Impact of natural organic matter and divalent cations on the stability of aqueous nanoparticles, *Water Res.*, *43*, 4249–4257.

9 Tailoring Microalgae for Efficient Biofuel Production

Hiren K. Patel and Jaydeep Dobariya
School of Science, P. P. Savani University, Surat, Gujarat, India

Rishee K. Kalaria
Aspee Shakilam Biotechnology Institute, Navsari Agricultural University, Surat, Gujarat, India

CONTENTS

DOI: 10.1201/9781003165101-9

9.1 INTRODUCTION

Seaweeds (sea macroalgae) are microscopic, multicellular, benthic marine algae that look like plants. The term comprises red, brown, and green algae, which are classified according to the thallus color derived from their dominant pigments (phycoerythrin and phycocyanin in red algae, chlorophyll a and b in green algae, and fucoxanthin in brown algae). They generally live, attached to hard substrates (such as rocks) in coastal areas, although some brown algae in laminariales and red algae in corallinales can live at depths of several or occasionally nearly a hundred meters below the sea surface (Lüning, 1990). Several species/populations (Sargassum and Ulva) have evolved to be free-floating by modifying their intercellular gas sacs to retain their desired depth in the water. Seaweeds are a polyphyletic group with no biologically common multicellular parent. There are approximately 10,000 species of seaweed worldwide. They originated through multiple endosymbiotic events during geological time. Generally speaking, green and red algae originated from a primary endosymbiosis when a eukaryotic host cell acquired an ancestral cyanobacterium as its plastid to form a primary symbiotic oxygenic eukaryote (Falkowski et al., 2004). Brown algae derived from a secondary endosymbiosis whose ancestor historically possessed a cryptic green algal endosymbiont that was subsequently replaced by a red algal chloroplast (Dorrell and Smith, 2011). Due to their genetically polyphyletic origin, they are now classified in different kingdoms (brown algae are in the kingdom Chromista, green algae and red algae are in the kingdom Plantae) (Cavalier-Smith, 1981), although all are called seaweed in the assemblage.

Unlike higher plants, seaweeds, most of which display an alternation of generation, have far more complicated life histories due to the flexible relationship among morphological phases, cytological events, and genetic behaviors than generally realized (West and Hommersand, 1981). Many red algae have three generations, two sporophytes (diploid) generations, the carposporophyte, and tetrasporophyte, in addition to a gametophyte (haploid) (West and Hommersand, 1981), most of which are morphologically macroscopic (e.g., adult thallus in Porphyra). The gametophyte and tetrasporophyte phases are usually morphologically similar (Kohlmeyer, 1975), although they have markedly different physiological behaviors (von Stosch, 1965). In some species, the carposporophyte is absent in sexual life history, and the male and female gametophytes are vegetatively dimorphic (van der Meer and Todd, 1980). Brown algae have heteromorphic (saccorhizapolyschides, laminaria), monophasic (compsonema saxicola), and isomorphic (ectocarpussiliculosus) life histories, with a lot of variation between them (Pedersen, 1981). The genus laminaria has a macroscopic diploid generation that creates sporangium, after which the cells divide into haploid zoospores by meiosis before being released and developing to microscopic male and female gametophytes (Tseng, 1987). In the life cycle of green algae, sexual reproduction may be isogamous, anisogamous, or oogamous (Lee, 2008). Ulva alternates between macroscopic isomorphic diploid and haploid phases; however, in some animals, there may be as many as 12 morphologically similar phases with cytological and reproductive variations (Tanner, 1981). The wide variety of life histories that seaweeds have, from haploid dominance to diploid dominance, is one of their most crucial differences (Otto and Gerstein, 2008).

9.2 GENETIC ENGINEERING

Genome engineering is focused on the development of methods to precisely manipulate nucleic acids in living cells. Targeted genome modification in plants has been regarded as an elusive goal (Shukla et al., 2009). However, recent advances in sequence-specific genome-engineering technologies have enabled the control of genetic material via targeted genome modifications (Voytas, 2013) in model plant organisms. These tools can be grouped into two categories: protein-directed and nucleotide-directed specificities (Esvelt and Wang, 2013). Zinc-finger nucleases (ZFN) have been successfully applied in chlamydomonas (Sizova et al., 2013). However, ZFN technology suffers from difficulties in design, construction, cost, and uncertain success rates (Jiang et al., 2013; Mahfouz et al., 2011). We'll concentrate on two recently established controlled genome-modification methods in this study. TALEs or TALs are protein-directed transcription activator-like effectors, and RNA-directed type II prokaryotic clustered regularly interspaced short palindromic repeats are RNA-directed type II prokaryotic clustered regularly interspaced short palindromic repeats (CRISPR). For other target genomic-engineering tools such as ZFN, readers may refer to several excellent and comprehensive reviews summarizing recent advances in precise genome-editing technology (Voytas, 2013; Esvelt and Wang, 2013; Gaj et al., 2013; Liu et al., 2013).

TALs (Moscou and Bogdanove, 2009) have rapidly developed and been utilized to create site-specific gene-editing (Li et al., 2012). TALs are proteins produced by the pathogenic plant bacteria, xanthomonas, when they infect plants through the typeIII secretion pathway (Bogdanove et al., 2010). These proteins can activate the expression of plant genes by binding effector-specific DNA sequences through their central tandem repeat region where the 12th and 13th amino acid residue corresponds to a specific nucleotide sequence (Bogdanove and Voytas, 2011) and transcriptionally activates gene expression. TAL is less costly, has fewer design requirements, and has fewer off-target behaviors than ZFN (Mussolino et al., 2011). In contrast to ZFN, it is easier to engineer new DNA-binding specificities (Christian et al., 2010). It has been used in model plants to change the expression of reporter genes in tobacco (Mahfouz et al., 2011; Zhang et al., 2013], insertions and deletions in arabidopsis thaliana, gene knockouts in rice (Cermak et al., 2011), and the generation of disease-resistant rice (Li et al., 2012). There also have been calls for this technology to be applied to microalgae homologous recombination (Weeks, 2011). Several in-house methods of synthesizing TAL proteins to target specific DNA sequences in chlamydomonas have been established, fused with Fok-I nuclease. This technology could be used to create specific modifications in the chlamydomonas genome (Borchers et al., 2014; Brueggeman, 2013).

CRISPR has been hailed as a revolution in genomic engineering (Rusk, 2014). It's a type of bacterial and archaeal immune system found in 40% of eubacteria and 90% of archaea (Horvath and Barrangou, 2010). It consists of three core components: RNA-guided CRISPR associate protein (Cas9) nuclease, CRISPR RNA (crRNA) and trans-acting crRNA (tracrRNA) (Xu et al., 2014), although occasionally the latter two components can be fused into a single guide RNA (sgRNA) (Jinek et al., 2012). CRISPR can directly edit the genome by either non-homologous ending joining (NHEJ), producing undefined indels, or template-dependent homology-directed repair (HDR), which leads to a defined DNA substitution, deletion, and insertion (Xu et al., 2014). In arabidopsis, the efficiency of HDR-mediated insertion is higher than NHEJ-mediated insertions (Li et al., 2013). When CRISPR fuses with effectors or transcriptional repressor domains, it generates stable and efficient transcriptional repression or activation or robustly silences multiple endogenous gene expression in human and yeast cells (Gilbert et al., 2013). This system has been successfully used in plant organisms to perform gene modification and mutagenesis (Jiang et al., 2003; Nekrasov et al., 2013), multiplex gene editing (Li et al., 2013; Shan et al., 2013), genome-scale screening (Shalem et al., 2014; Wang et al., 2014), transcriptional control (up-regulated and down-regulated) (Gilbert et al., 2013), dynamic imaging of chromosome activity (Chen et al., 2013), and even multiplexed RNA-guided transcriptional activation, repression, and gene editing simultaneously. Presently, the majority of studies using CRISPR engineering that has been related to genome screening and transcriptional regulation have been performed in bacterial or animal cells (Xu et al., 2014).

Realistically, both the TAL and CRISPR techniques are not mature enough to be applied universally. Efforts must be made to resolve many fundamental problems in both methods; for example, the molecular structure and catalytic mechanism of the CRISPR complex (Xu et al., 2014) and the pathway of TALs delivery into cells by lentiviruses. Additionally, studies need to be performed for these targeted genome-

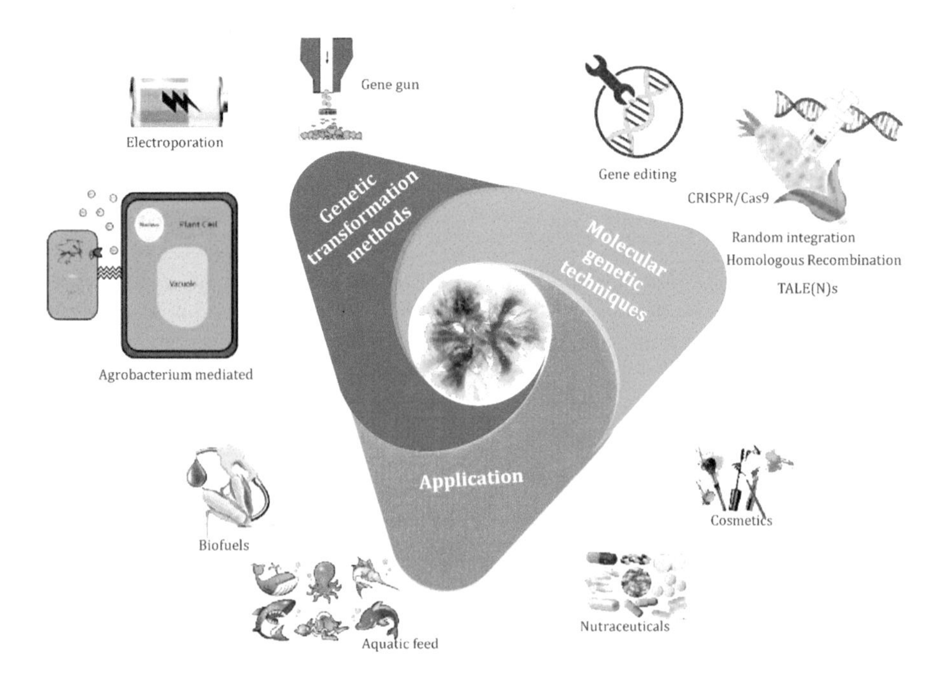

FIGURE 9.1 Recent developments on genetic engineering of microalgae for biofuels.

editing tools to determine their safety and specificity to decrease off-target possibilities (Fu et al., 2013; Podevin et al., 2012) and to compare their efficiencies. While the aforementioned groundbreaking experiments in algae have mainly been performed in the model microalga chlamydomonas, it is not too early to begin genetic-modification studies in seaweed. Based on the assumption that closely phylogenetically related species may share similar genetic, biochemical and physiological, and morphological features (Keeling et al., 2005). Research in unicellular microalgae including porphyridium purpureum in rhodophyta and thalassiosira pseudonana in heterokontophyta may have contributed to the development of genetic-modification systems (Bhattacharya et al., 2013) (Figure 9.1).

9.3 WHY NEED GENETIC ENGINEERING OF SEAWEED?

Marine macroalgae are a complex group of organisms with varying morphologies and bioactive properties. The capacity for genetic improvement to increase the value of such molecules and other economically significant characteristics is likely to be high because procedures for cultivating several organisms are well known, genetic diversity is often high, sexual proliferation is often possible, and seaweeds typically have a short generation period that enables rapid transfer from one selected generation to the next one. Seaweed gene editing is a type of genetic-modification expression mechanism that differs from that found in heterotrophic prokaryotes and higher plants. Seaweed genetic engineering has the potential to bridge the gap between fundamental and applied seaweed science. The reducing cost of sequencing helps us to examine the fine genetic structure of seaweeds and, as a result, recognize novel promoters, transformation approaches, selective markers, and so on. These breakthroughs will open up new opportunities for seaweed genetic engineering, which is also in its early stages. Combining these developments with the increasingly evolving fields of systems biology and metabolic engineering would fulfill energy, environmental, and human health demands.

9.4 SEAWEED GENOMICS AND MODEL ORGANISM SELECTION IN SEAWEED

Microalgae have a lot of genomics data, and it's very complete, whereas seaweed has a lot of genomics data, but it's very limited. Ectocarpussiliculosus, a brown algae closely related to laminaria spp., has a genome that sheds light on the physiology and evolution of multicellularity in brown algae. In this genome, extended sets of light-harvesting and pigment biosynthesis genes and new halide metabolic processes have been discovered (Cock et al., 2010). For carbon storage, the central pathways of carbohydrate and protein glycosylation are well conserved, while a complicated laminar in metabolism replaces glycogen and starch metabolism from the secondary endosymbiont (Michel et al., 2010). The first full nuclear genome of red seaweed has been discovered. Despite its large size (105 megabase pairs [Mbp]), chondrus crispus has a small genome with just a few functional gene duplications. The chondrus genome, like the ectocarpus genome, has a halogen metabolism mechanisms that helps it adapt to the tidal coastal climate. Chondrus' carbohydrate metabolism indicates that cellulose synthesis is polyphyly, and that red algae's mannosylgly cerate synthase may have evolved from a marine bacterium. Red algae genomes have experienced loss and expansion throughout evolutionary history, including the loss of genes, introns, and intergenic DNA due to ecological factors, followed by an expansion of genetic background due to transposable element activity (Collen et al., 2013). The genome of pyropiayezoensis, one of the most commercialized and well-cultivated seaweeds, has also been sequenced. In its 43Mbp genome, 35% of the genes are functionally uncharacterized, and a second homolog of the phycobilisome-degradation gene, which had been assumed to be chloroplast derived, was found in the nuclear genome. This newly discovered gene may be involved either in phycobiliso mephoto bleaching or in P. yezoensis nitrate metabolism (Nakamura et al., 2013). With the significant development in next-generation sequencing(NGS) technologies, the cost efficiency has decreased and the volume of genome sequencing has increased (Mardis, 2011; Mardis, 2013). The whole-genome sequencing of saccharina japonica, an economically important seaweed with a genome size of 580–720 Mbp, was recently announced (Legall et al., 1993), has been completed and the total number of genesis estimated to be upto 35,725; larger than any other eukaryotic algae (Yao and Jun, 2014). Furthermore, basic genomic information for seaweeds with large genomes is being provided (Cock et al., 2010; Collen et al., 2013) and has a symbiotic bacteria that produces noise (Nakamura et al., 2013), NGS's lower cost and improved sequencing productivity and effectiveness enable researchers to look at species or strains other than model organisms, opening up new possibilities for functional genomics within the same phylogenetic seaweed group.

In macrocystispyrifera, transcriptomic analysis assessed gene expression in response to various abiotic factors such as light, temperature, and nutrients, and revealed new gene families in brown algae. The assembly of the 228 Mbp sequence revealed high genetic similarity between macrocystispyrifera and its brown seaweed relatives ectocarpus and laminaria (Konotchick et al., 2013). These breakthroughs may produce complete seaweed genomes that could shed light on physiology, ecology, reproduction, evolution, etc., which are essential in genetic engineering.

9.5 GENETIC MANIPULATION AND IMPROVEMENT IN SEAWEED

9.5.1 Soma Clonal Variants

Seaweed cellular biotechnology started in the 1980s and is still well behind that of terrestrial plants. The creation of in vitro culture systems in macroalgae allows for year-round mass proliferation of biomass for the production of useful compounds under defined conditions. Novel genetic variants with useful characteristics resulting from somaclonal variation, can also be found in culture systems. In seaweeds, techniques for somatic embryogenesis and dedifferentiation of somatic cells often result in morphological variations. Such phenotypic and physiological variation can be useful to genetic improvement projects and can also be used to store and pick germplasm. The new change could be temporary, reversible, or permanent. Temporary changes are mostly caused by epigenetic or physiological changes, all of which are reversible, even after a few years (Kaeppler et al., 2000). Permanent somaclonal variants' underlying molecular processes are scarcely studied and incompletely known (Larkin and Scowcroft, 1981). Nonetheless, the majority of morphological variation found during seaweed in vitro culture is temporary and not passed on to offspring. There are several cases of the same genotype developing morphological and anatomical phenotypes. For example: in vitro tissue culture of kelp (order laminariales) sporophytes have a frequent developmental pattern in which outgrowths of aposporous gametophyte-like filaments with differentiated fertile branches can give rise directly to sporophytes (Ar Gall et al., 1996). Similarly, meristem cultures of laminari are generate into one of three different body types: (1) uniseriate filaments; (2) thalloid-like structures; (3) dark green, compact calli. Similarly, the early development of embryonic germlings from 22 fucaceae species can show up to six developmental forms.

Plant protoplasts can be genetically modified in vitro, allowing for the production of genetically better agricultural crop varieties. Various investigations on the isolation and regeneration of protoplasts from a broad range of seaweed body types, from simple leafy thalli to complex, cylindrical, branched thalli, have been conducted. Seaweed protoplasts, unlike higher plants, replicate and distinguish into a full thallus without the need for phytohormones in the growth media. Green seaweed protoplasts, on the other hand, have a range of morphogenetic variations and give rise to a variety of morphologically complex morphotypes, including free-living sporangia, micro-thalli, or saccate (or spherical), tubular (or spindle), irregular, or frondose thalli with varying life spans. Three distinct protoplast regeneration processes have been established in the red alga porphyra: callus form, (2) filamentous form, and (3) conchocelis form (Polne-Fuller et al., 1984). From protoplast culture of numerous brown macroalgal species, callus-like outgrowth (i.e. an unorganised cellular mass) has been observed. Similarly, two distinct forms of protoplast-based growth have been identified in gracilariagracilis, resulting in plants with different appearances and lifespans. Regenerated plants have either slender, branched thalli like parental plants or remain small with dense, unbranched thalli, many of which die. Despite the lack of understanding of the underlying processes, this variance in developmental morphological anomalies has been due to a number of factors, including the type of donor tissue used to prepare protoplasts and the culture conditions used for regeneration. Protoplasts extracted from monos- Troma altissimo's vegetative thalli, for example, regenerate into regular thalli, while protoplasts isolated from the holdfast grow into filaments (Chen, 1998). Ulvafasciata protoplasts mature into microthalli when cultured in high density (Chen and Shih, 2000). Protoplast regeneration into normal sporophytes can occur in brown seaweeds of the order laminariales through a variety of developmental processes, including direct regeneration into plantlets, or indirect regeneration, depending on water temperature in some species. Indirect regeneration occurs either after dedifferentiation of the tissue into a filament (U. pinnatifida; Matsumura et al., 2001, L. saccha-rina; Benet et al., 1997) or after the development of callus-like masses (U. pinnatifida; Matsumura et al., 2001) and L. japonica; Matsumura et al., 2000). Gupta et al. (2012) provide the first report of epigenetic regulation of morphology in protoplast-derived germ lings, with DNA methylation acting as an underlying molecular mechanism in protoplast-derived germ lings in ulva reticulata. The production of irregular thalli may also be caused by axenic culture conditions. The majority of studies have found that seaweed-associated microflora produces morphogenetic substances that result in normal thallus structure (Matsuo et al., 2005; Spoerner et al., 2012). However, due to economic and farming advantages, most commercial seaweed production is currently focused on simple vegetative propagation. For high-end applications, in vitro culture techniques for seaweeds are being developed to generate new genetic variants or facilitate clonal propagation in photobioreactors.

9.6 INDUCING MORPHOLOGICAL VARIATION THROUGH SOMATIC HYBRIDIZATION

Hybridization, which combines phylogenetically distinct genetic lineages and results in morphotypes that are either intermediate to the parental species or entirely novel, is an important process that occurs in addition to the morphological variation caused by in vitro culture. Species are not always correctly described in seaweeds, and they are constantly being delineated using molecular genotyping or sequencing (Maggs et al., 2007). Hybridization may bring together divergent genetic lineages by crossing two genetically distinct populations (Hodge et al., 2010). The resulting hybrid blends parental phenotypes or creates a new phenotype that is distinct from the parental form. Somatic hybridization through protoplast fusion, in contrast to environmental hybridization events that lead to speciation, holds great promise for achieving large crosses between species that are hard or impossible to transfect traditionally (Davey et al., 2005). Fusing protoplasts from different backgrounds takes advantage of natural genetic variation and creates new genetic combinations that could enhance a variety of functional traits. Protoplast fusion, on the other hand, results in either heterokaryon (the fusion of nuclei from different species) or homokaryon (the fusion of nuclei from the same species) (fusion of nuclei of same species) the mechanism of recombination. The increase in ploidy will boost agronomic traits in both cases. To differentiate heterokaryons from homokaryons and unfused parental protoplasts, most studies use color differences. Somatic hybridization has achieved unprecedented success in terrestrial plants, but only a few attempts in seaweeds have been developed. Furthermore, almost all protoplast fusion studies conducted to date lack a detailed explanation of cell division and developmental stages. Below is a description of the regeneration of fusion products recorded so far from various seaweed species.

9.7 INTERGENERIC HYBRIDIZATION

Ulvapertusa and enteromorphaprolifera strains of ulvapertusa and enteromorphaprolifera have been used in subsequent research on intergeneric protoplast fusion (Reddy et al., 1989). Because of their larger size and the existence of twin chloroplasts, presumptive heteroplasmic fusion products were defined. Following that, study of the regeneration patterns of fused protoplasts indicated that they were close to the formation of regular (unfused) protoplasts. The thallus of most regenerated plants from fusion products resembled either U. pertusa or E. prolifera. The thalli of some plants, on the other hand, had a distinctive irregular and dentate margin that was never seen in the parental form.

9.8 TRANS DIVISIONAL HYBRIDIZATION

Kito et al. (1998) published the first report of successful trans divisional protoplast fusion between monostroma and porphyra. The regeneration processes and characteristics of protoplast fusion products from these two organisms were very different. Although the fusion partners displayed distinct monostromatic and distromaticthalli, the regenerated hybrids were green with a distromatic structure. Initial heterokaryons were identified based on clearly distinguishable chloroplast colors but were indistinguishable after 5 days of culture. Of the hybrids generated, one of the heterofusant plants grew into a multicellular body followed by the development to rhizoid-like and bud-like organs. Another hybrid developed a multicellular body that, after further cultivation, divided into individual cells, each of which grew into a long stringy plant; this mutant was given the name "kattsunbo." A bud grew from a third fusion product. This bud-like plant grew into a thallus (7.5 cm long and 6.5 cm wide) and was given the name "nigo." These few examples show the wide range of morphology that somatic hybridization in seaweeds can produce. Besides that, methods have been developed to isolate seaweed protoplasts and bring them to complete thallus regeneration on a variety of seaweeds, along with the most anatomically complex taxa like laminaria, undaria, gracilaria, and kappaphycus, which are among the most economically available seaweed genera. Nonetheless, somatichybrids have not yet been widely established, and no "cultivars" for field cultivation have been made. However, the progress made so far in the creation of homo- or hetero-karyons by somatic hybridization is useful preparation.

9.9 MUTAGENESIS MEDIATED MORPHOLOGICAL VARIATION

Macroalgal mutants have been little studied. In 1958, Ralph Lewin emphasized the importance of macroalgal genetics(Lewin, 1958); now, more than 50 years later, still very few mutants have been isolated and analyzed. In addition to the difficulties of growing macroalgae in laboratory conditions and their complicated and relatively long life cycles, there is a surprising lack of interest in this field of study. As a result, and despite the amenability of haploid organisms for genetic and molecular analyses of underlying mechanisms and pathway, the identification of genes involved in the control of marine macroalgal growth and development lags far behind that of other studied multicellular organisms (Howell, 1998).

The most advanced genetic characterization of developmental mutants in a multicellular alga has been carried out in the chlorophyceae taxon volvoxcarteri (subdivision chlorophyta; Matsuzaki et al., 2012). This microscopic freshwater alga is made up of around 2,000 biflagellated cells that form a moving spherical body inside a thick extracellular gelatinous matrix. Although the majority of cells remain flagellated and vegetative, asymmetric cell divisions turn 16 cells into non-flagellated, larger asexual reproductive cells (gonidia) (Starr, 1969). Following this, the embryo inverts, the flagellated somatic cells are externalised, and the gonidia are internalised.

Chemical mutagenesis and transposon tagging in V. carteri resulted in a number of mutants with morphological defects (Sessoms and Huskey, 1973). The glsA gene, which codes for a chaperone protein that regulates both protein translation and transcription (Miller and Kirk, 1999; Pappas and Miller, 2009), is responsible for the asymmetric cell division that gives rise to the gonidia cells. Additional mutations led to the discovery of transcription factors that regulate genes unique to either gonidial cells (lag gene) or somatic cells (somatic gene) (regA gene) (Kirk, 2003). In addition, the invB mutant has been shown to code for a nucleotide-sugar transporter that is needed for the expansion of the glycoprotein-rich gonidia vesicle, which tightly surrounds the multicellular sphere before inversion (Ueki and Nishii, 2009).

9.10 TRANS-CONJUGATION

It is the transfer of the DNA between a cyanobacterial cell and bacterial cell. It is usually done by direct cell-to-cell contact or bridge-like connection between two cells. It is a versatile method that can be used for genetic manipulation of marine and freshwater algae. This method is demonstrated first to trans-conjugation of 5 strains of marine cyanobacteria synechococcus, synechocytis, and pseudanabaena. Conjugation was done using transposon and vector pKT230 (IncQ). This research confirmed wide application of conjugation in microalgae. A plasmid containing green fluorescent protein (GFP) was transferred into prochloroc occus strain by interspecific conjugation with E.coli and protein expression was detected by Wester blotting and cellular fluorescence (Tolonen et al., 2006).

9.11 NATURAL TRANSFORMATION AND DIRECT INDUCED TRANSFORMATION

This method allows to absorb extracellular DNA directly in naturally competent cells or artificially induced competent cells. In marine algae, natural transformation has been reported for only synechococcus sp. PCC7002, and others are mostly freshwater strains of cyanobacteria. It uses treatment with ethidium bromide for the strain (Qin et al., 2012). The mechanism of competence is almost similar in both type of cyanobacteria, but transformation efficiency of marine synechococcus were much lower than freshwater synechococcus strains. It happens mainly due to the presence of some polysaccharides, which hinder the uptake of DNA. Nowadays, quick and simple transformation methods such as electroporation are available for marine strains.

9.12 ELECTROPORATION

Electroporation method is used for transformation in bacterial cells for a long time. It is a simple and highly efficient method for a small amount of DNA (Neumann et al., 1982). By the electroporation technique, extrinsic DNA can be transferred independent from a cell's ability. Electroporation is first performed in marine cyanobacterium synechococcus sp. (Matsunaga et al., 1990). The strength of an electric field required for marine cyanobacteria was than that of freshwater strains. This efficiency can be increased by pretreatment with CaCl2. Electroporation-mediated transformation was also achieved in eukaryotic chlamydomonas reinhardtii strains using 14Kb plasmid (Brown et al., 1991). The electroporation method is much more efficient than the glass beads method to transfer exogenous DNA. Lately, the transformation protocol has been established for the oil-producing algae nannochloropsis sp. by the electroporation method, and several genes were knocked out using the homologous recombination method (Kilian et al., 2011).

9.13 BIOLISTIC TRANSFORMATION

The microparticle bombardment method has been the most efficient method for direct gene transfer. There were protocols established for the transformation of many nuclear and chloroplast expression systems of microalgae (Qin et al., 2012).

Advantages of biolistic method: -

- Including plants, animals, microbes, various cells and tissues can be introduced by exogenous DNA by this method, by biolistic method chloroplasts, mitochondria and other organelles.
- Due to limited information of genomes of most algae, it is difficult to design endogenous vector. Vectors from E.coli. were usually helpful in algal biolistic transformation.
- Gene gun usually used for particle bombardment, but it can be controlled and mostly all physical and chemical parameter can be adjusted.

Marine cyanobacterium synechococcus was transformed using particle bombardment with bacterial magnetic particles. The particle is covered with a phospholipid layer and that can bind larger quantities of DNA (Matsunaga et al., 1990). Tools available for genetic manipulation of diatoms is limited. Biolistic transformation is the most efficient tool for transformation in diatoms. Transformation methods with particle bombardment have been established for many species of diatoms, including thalassiosira psuedonana, thalassiosira weissflogii, C. cryptica (Qin et al., 2012).

9.14 MICROINJECTION

It is direct physical method to penetrate the cell wall. It is does not require a protoplast regeneration system. Microinjection is helpful to introduce substances in a very controlled manner and at specific targets. Due to difficulty in immobilizing algal cells, the microinjection method is rarely used in marine algae transformation (Neuhaus and Spangenberg 1990). A high-yield transformation was established for marine green algae acetobularia mediterranea by microinjection (Neuhaus et al., 1986). Nevertheless, it is a complicated and delicate procedure, and it can be considered to be highly efficient and lost-cost transformation method for marine algae.

9.15 ARTIFICIAL TRANSPOSON METHOD

Transposons are mobile DNA elements discovered in maize. They are used as a genetic tool for genetic manipulation. Artificial transposons can separate and transform from natural one. It has been developed for in vitro mutagenesis and genetic transformation. The high frequency of transposition by artificial transposons, which can integrate foreign DNA into receptor cell's genome, which avoid random integration of genes (Wu et al., 2011). Using natural Tn5, transpose and cation liposomes are complexed with the help of electroporation, which improve the transformation efficiency (Reznikoff, 2008).

9.16 AGROBACTERIUM TUMEFACIENS-MEDIATED GENETIC TRANSFORMATION

Agrobacterium tumefaciens-mediated transformation transforms plants by transferring and integrating Ti plasmid with large DNA fragment to plant genome with the help of Vir proteins for T-DNA transfer. In marine seaweed porphyrayezoensis, genetic transformation by A.tumefaciens first reported. The transformation frequency of gene transfer to freshwater alga C. reinhardtii by A. tumefaciens freshwater strains is much higher than that of the glass-bead transformation (Kumar et al., 2004). Transformation by this method is affected by many factors, such as strains used and plasmid vectors. Due to low copy numbers of Ti plasmid, plasmid isolation and manipulation become more difficult, and hence, transformation becomes more difficult.

9.17 GENETIC ENGINEERING OF OTHER ALGAE FOR BIOFUEL PRODUCTION

While the molecular toolkit of C. reinhardtii is highly developed, it seems unlikely that this freshwater green alga will be a commercial biofuel producer. Prior to the recent investigation of microalgae for biofuels, only a few species have been genetically modified outside of chlamydomonas. These include other green algae, diatoms, dinoflagellates, and red algae. Microparticle bombardment, electroporation, and agitation with silicon whiskers are all effective transformation methods for C reinhardtii. The majority of species that have been transformed began with the complementation of a recognized mutation or the introduction of a resistance marker as the first step. These methods and vectors, which are discussed further below, are useful tools that should enable these species to undergo further genetic and metabolic engineering.

9.17.1 Green Algae

In addition to C. reinhardtii, nuclear transformations of several species of green algae have been reported. The primary strategy for transformation of V. carteri is through particle bombardment. Haematococcus pluvialis is an important producer of the keto-carotenoid as taxanth in and has been transiently and stably transformed by particle bombardment (Steinbrenner and Sandmann, 2006b). Particle bombardment has effectively transformed the halotolerant green microalga dunaliella-salina, which is used to manufacture b-carotene (Tan et al., 2005), electroporation (Geng et al., 2004; Sun et al., 2005), and agitation with glass beads (Feng et al., 2008). Particle bombardment and protoplast electroporation are the primary methods for transformation of most chlorella species; however, initial attempts only generated transient transformants. Furthermore, evidence suggests that nuclear transformation of C. sorokiniana proceeds through homologous recombination (Dawson et al., 1997). Finally, microparticle bombardment was used to achieve stable nuclear transformation of the volvocine alga gonium pectoral (Lerche and Hallmann, 2009). The heterologous genes aph VIII and luciferase were expressed under the control of C. reinhardtiior V. carteri promoters and UTRs.

9.17.2 Diatoms

Diatoms are a large and diverse group of unicellular eukaryoticalgae known for their distinctive silica-based cell walls, frustules. Diatoms are an essential source of energy, as well as CO_2 fixation and O^2 processing, in the oceans. They're also useful commercial sources of aquaculture feed and specialty oils including omega-3 fatty acids (Apt and Behrens, 1999). Furthermore, there has recently been a wave of interest in discovering new applications for diatoms in nanotechnology (Davis and Hildebrand, 2010).

9.17.3 Dino Flagellates

Dino flagellates are a diverse group of eukaryotic algae that live in both freshwater and marine environments. Around half of them are photosynthetic, and the others are unicellular. There has only been one study of dinoflagellate nuclear transformation to date (Te, Lohuis, and Miller, 1998). Three different bacterial genes were introduced, nptII (G418resistance), hpt (hygromycin B resistance), and the reporter gene b-glucuronidase (GUS) into two dinoflagellates, amphidinium sp. and symbiodinium microadriaticum, by agitation with silicon whiskers (Te, Lohuis, and Miller, 1998). Each gene was fused to a heterologous promoter and appeared to be incorporated into the nuclear genome in a stable manner.

9.17.4 Red Algae

There have been several reports on the transformation of red algae: plastid transformation of porphyridium spp. Bymicroparticle bombardment (Lapidot et al., 2002), nuclear transformation of the red macroalgae por-phyrayezoensis by agrobacterium-mediated gene transfer (Chen et al., 2001), and nuclear transformation of cyanidioschyzonmerolae (Minoda et al., 2004) (Figure 9.2).

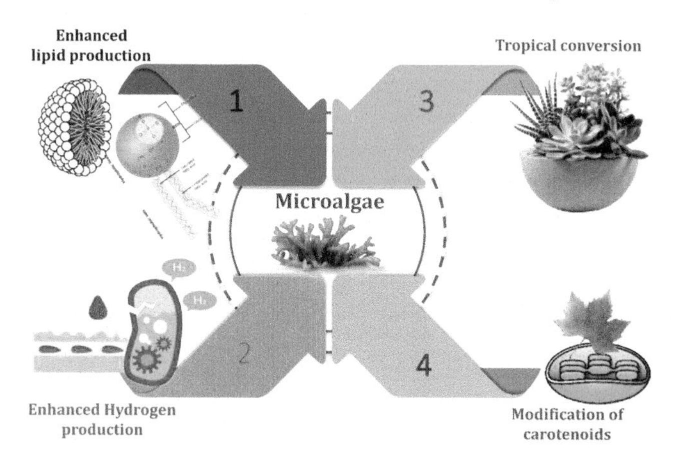

FIGURE 9.2 Metabolic engineering to improve characteristics of microalgae.

9.18 METABOLIC ENGINEERING OF MICROALGAE

Despite the growing popularity of algae as a biofuel production platform, few studies have shown that genetic engineering can be used to enhance biofuel strains. However, as discussed below, significant genetic-engineering attempts have been made for strain improvement, biohydrogen production, and metabolic engineering for carotenoid production. Furthermore, because of the potential for industrial applications, it is likely that much of the continued research in the lipid-based biofuel area has yet to be released.

9.18.1 Enhanced Lipid Production

At the time of this writing, only one report directly attempts to increase lipid production in microalgae. For this purpose, the acetyl-CoA carboxylase (ACCase) gene, which is the enzyme that catalyzes the first committed step for fatty acid synthesis, was cloned from the diatom cyclotella cryptica (Roessler and Ohlrogge, 1993). Expression vectors were made, and transformation protocols were developed for C. cryptica and the diatom naviculasaprophila. The authors reported a two- to threefold increase in the level of ACCase activity for the transformed diatoms, but no increase in fatty acid accumulation was detected. However, no experimental data was presented for the increase of ACCase activity (Dunahay et al., 1996). Furthermore, metabolic engineering of E. coli and plants has shown that overexpression of ACCase alone is not a good strategy, probably because it might not be a limiting step or there is an additional bottleneck in the pathway (Courchesne et al., 2009). Despite the fact that this method did not achieve the desired results, it was a great proof of concept that a metabolic-engineering framework could be designed from the ground up for a particular microalgae strain.

9.18.2 Enhanced Hydrogen Production

Hydrogen has also been studied as an alternative for algae-derived fuels (Mehta et al., 2009). C. reinhardtii, chlamydomonas moewusii, scenedesmusobliquus, lobochlamysculleus and chlorella fusca have been shown to produce hydrogen under anaerobic conditions in the presence of light. However, this method is inefficient and difficult to replicate on a large scale (Ghirardi, 2000; Meuser et al., 2009). Both of these steps may be enhanced by genetic modification, including hydrogenase oxygen sensitivity, anaerobiosis induction, avoiding competition for electrons from other pathways, and increasing electron sources (Beer et al., 2009; Melis, 2007).

9.18.3 Trophic Conversion

Growing microalgae under heterotrophic or mixotrophic conditions via the addition of a reduced carbon source has several advantages. Fermentation systems have been extensively researched and implemented in the industry. The culture conditions are highly controlled and reproducible. Furthermore, heterotrophic microalgae cultures reach higher cell densities, resulting in lower harvesting costs (Chen and Chen,

2006). Furthermore, compared to autotrophic cultures, heterotrophic cultivation of chlorella proto thecoides for biofuel production resulted in a fourfold increase in lipid accumulation (Xu et al., 2006).

9.19 METABOLIC ENGINEERING OF CAROTENOIDS

Dunaliellasalina and haematococcus pluvialis are the main sources for commercial production of beta-carotene and as taxanthin, respectively. Other microalgae also show great potential for producing other relevant carotenoids such as lutein (Dufossé et al., 2005). Despite this, few reports on metabolic engineering for increased carotenoid production in algae have been published. Wong attempted to produce keto carotenoids (e.g. astaxanthin) in C. reinhardtii by overexpressing beta-carotene ketolases from H. pluvialis (bkt3) and C. reinhardtii itself (CRBKT) (Ngan et al., 2006). They could detect small accumulation of 4-ketolutein, which was not present in the parental strain.

9.20 CONCLUSION

Though the genetic engineering of microalgae holds great potential to improve process economics, it is limited mainly due to the unavailability of the genetic information for robust and commercially suitable strains. In recent times, rapid advances in DNA synthesis, genetic manipulation tools and techniques, and availability of functional enomes have improved the chances to better engineer microalgae with complex functions. However, the information on genetic-strain design criteria continues to stymie development. Furthermore, once the genetically engineered strains are produced, their commercial success will be determined by their protection to human health and the environment. Therefore, it is recommended that strict regulations and monitoring should be in place to evaluate the environmental and human health risk of using GM microalgae particularly in outdoor cultivation. Here, the recent development in precise genome-editing technologies such as nontransgenic and marker-free CRISPR has the potential to revolutionize the microalgal bioengineering for the production of non-GMO algal products. The non-GMO tag to the bioengineered microalgae is expected to improve the biosafety and alleviate the regulatory issues associated with the usage of GM microalgae.

Seaweed genetic engineering could bridge the gap between fundamental and applied studies in seaweed research. The decreasing cost of sequencing provides us with many opportunities to investigate the fine genetic structure of seaweeds and consequently identify innovative genetic-transformation elements (promoters, transformation methods, selective markers, etc.). When it comes to environmental concerns, multidisciplinary research on algal genetics, physiology, reproductive biology, and ecology is needed, from the molecular to at least the local aquatic ecosystem level, using mathematics, biology, chemistry, physics, and even sociology. These innovations will bring unlimited possibilities in seaweed genetic engineering, which is still in its infancy. Combining these innovations with the rapidly developing fields of systems biology and metabolic engineering will satisfy the demands related to energy, environment, and human health.

REFERENCES

Apt, Kirk E., and Behrens, Paul W. (1999). Commercial developments in microalgal biotechnology. *Journal of Phycology*, 35, 215–226.

Ar Gall, E., Asensi, A., Marie, D., and Kloareg, B. (1996). Parthenogenesis and apospory in the Laminariales: A flow cytometry analysis. *Eur. J. Phycol.*, *31*, 369–380.

Beer, Laura L, Boyd, Eric S, Peters, John W, and Posewitz, Matthew C (2009). Engineering algae for biohydrogen and biofuel production. *Current Opinion in Biotechnology*, 20, 264–271.

Benet, H., Gall, A.E., Asensi, A., and Kloareg, B. (1997). Protoplast regeneration from gametophytes and sporophytes of some species in the order Laminariales (Phaeophyceae). *Protoplasma*, *199*, 39–48. doi: 10.1007/BF02539804

Bhattacharya, D., Price, D.C., Chan, C.X., Qiu, H., Rose, N., Ball, S., Weber, A.P.M., Arias, M.C., Henrissat, B., Coutinho, P.M., et al. (2013). Genome of the red alga Porphyridium purpureum. *Nat. Commun.*, *4*(1), 1–10.

Bogdanove, A.J., Schornack, S., and Lahaye, T. (2010). TAL effectors: Finding plant genes for disease and defense. *Curr. Opin. Plant Biol.*, *13*, 394–401.

Bogdanove, A.J., and Voytas, D.F. (2011). TAL effectors: Customizable proteins for DNA targeting. *Science*, *333*, 1843–1846.

Borchers, A., Wright, D., and Spalding, M.H. (2012). Development of tal nucleases for genome modification in Chlamydomonas. Available online: http://www.cbirc.iastate.edu/files/2012/09/Development-of-TAL-Nucleases-for-Genome-Modification-in-Chlamydomonas.pdf (accessed on 25 March 2014).

Brown, L.E., Sprecher, S.L., and Keller, L.R. (1991). Introduction of exogenous DNA into chlamydomonas reinhardtii by electroporation. *Mol. Cell. Biol.*, *11*(4), 2328–2332.

Brueggeman, A.J. (2013). Transcriptomic Analyses of the CO_2-Concentrating Mechanisms and Development of Molecular Tools for ChlamydomonasReinhardtii. Ph.D. Thesis, University of Nebraska, Lincoln, NE, USA.

Cavalier-Smith, T. (1981). Eukaryote kingdoms: Seven or nine? *BioSystems*, *14*, 461–481.

Cermak, T., Doyle, E.L., Christian, M., Wang, L., Zhang, Y., Schmidt, C., Baller, J.A., Somia, N.V., Bogdanove, A.J., and Voytas, D.F. (2011). Efficient design and assembly of custom TALEN and other TAL effector-based constructs for DNA targeting. *Nucleic Acids Res.*, *39*, 7879.

Chen, B., Gilbert, L.A., Cimini, B.A., Schnitzbauer, J., Zhang, W., Li, G.-W., Park, J., Blackburn, E.H., Weissman, J.S., Qi,L.S., et al. (2013). Dynamic imaging of genomic loci in living human cells by an optimized CRISPR/Cas system. *Cell*, *155*, 1479–1491.

Chen, Guan-Qun, and Chen, Feng (2006). Growing phototrophic cells without light. *Biotechnology Letters*, 28, 607–616.

Chen, Ying, Wang, Yiqin, Sun, Yongru, Zhang, Liming, and Li, Wenbin (2001). Highly efficient expression of rabbit neutrophil peptide-1 gene in Chlorella ellipsoidea cells. *Current Genetics*, 39, 365–370.

Chen, Y.C. (1998). Development of protoplasts from holdfasts and vegetative thalli of Monostroma latissimum (Chlorophyta, Monostromatacae) for algal seed stock. *J. Phycol.*, *34*, 1075–1081. doi: 10.1046/j.1529-8817.1998.341075.x

Chen, Yean-Chang, and Shih, Hsiu-Chuan (2000). DEVELOPMENT OF PROTOPLASTS OF *ULVA FASCIATA* (CHLOROPHYTA) FOR ALGAL SEED STOCK. *Journal of Phycology*, 36, 608–615. doi:10.1046/j.1529-8817.2000.99128.x.

Christian, Michelle, Cermak, Tomas, Doyle, Erin L, Schmidt, Clarice, Zhang, Feng, Hummel, Aaron, Bogdanove, Adam J, and Voytas, Daniel F (2010). Targeting DNA Double-Strand Breaks with TAL Effector Nucleases. *Genetics*, 186, 757–761. doi:10.1534/genetics.110.120717.

Christian, M., Cermak, T., Doyle, E.L., Schmidt, C., Zhang, F., Hummel, A., Bogdanove, A.J., Voytas, Cock, J.M., Sterck, L., Rouze, P., Scornet, D., Allen, A.E., Amoutzias, G., Anthouard, V., Artiguenave, F., Aury, J.M., Badger, J.H., et al. (2010). The *Ectocarpus* genome and the independent evolution of multicellularity in brown algae. *Nature*, *465*, 617–621.

Cock, J.M., Sterck, L., Rouzé, P., Scornet, D., Allen, A.E., Amoutzias, G., … , and Wincker, Patrick (2010). The Ectocarpus genome and the independent evolution of multicellularity in brown algae. *Nature*, 465, 617–621 10.1038/nature09016.

Collen, J., Porcel, B., Carre, W., Ball, S.G., Chaparro, C., Tonon, T., Barbeyron, T., Michel, G., Noel, B., Valentin, K., et al. (2013). Genome structure and metabolic features in the red seaweed *Chondrus crispus* shed light on evolution of the Archaeplastida. *Proc. Natl. Acad. Sci. USA*, *110*, 5247–5252.

Courchesne, Noémie Manuelle Dorval, Parisien, Albert, Wang, Bei, and Lan, Christopher Q. (2009). Enhancement of lipid production using biochemical, genetic and transcription factor engineering approaches. *Journal of Biotechnology*, 141, 31–41.

Davey, M.R., Anthony, P., Power, J.B., and Lowe, K.C. (2005). Plant protoplasts: Status and biotechnological perspectives. *Biotechnol. Adv. 23*, 131–171. doi: 10.1016/j.biotechadv.2004.09.008.

Davis, Aubrey K., and Hildebrand, Mark (2010). Molecular Processes of Biosilicification in Diatoms. *Biomineralization*, pp. 255–294.

Dorrell, R.G., and Smith, A.G. (2011). Do red and green make brown?: Perspectives on plastid acquisitions within chromalveolates. *Eukaryot. Cell*, *10*, 856–868.

Dunahay, Terri G., Jarvis, Eric E., Dais, Sonja S., and Roessler, Paul G. (1996). Manipulation of microalgal lipid production using genetic engineering. *Seventeenth Symposium on Biotechnology for Fuels and Chemicals*, 57, 223–231.

Esvelt, K.M., and Wang, H.H. (2013). Genome-scale engineering for systems and synthetic biology. *Mol. Syst. Biol.*, *9*, 641.

Falkowski, P.G., Katz, M.E., Knoll, A.H., Quigg, A., Raven, J.A., Schofield, O., and Taylor, F.J.R. (2004). The evolution of modern eukaryotic phytoplankton. *Science*, *305*, 354–360.

Feng, Shuying, Xue, Lexun, Liu, Hongtao, and Lu, Pengju (2008). Improvement of efficiency of genetic transformation for Dunaliella salina by glass beads method. *Molecular Biology Reports*, 36, 1433–1439.

Fu, Y.F., Foden, J.A., Khayter, C., Maeder, M.L., Reyon, D., Joung, J.K., Sander, J.D. (2013). High-frequency off-target mutagenesis induced by CRISPR-cas nucleases in human cells. *Nat. Biotechnol.*, *31*, 822–826.

Gaj, T., Gersbach, C.A., and Barbas, C.F. (2013). ZFN, TALEN, and CRISPR/cas-based methods for genome engineering. *Trends Biotechnol.*, *31*, 397–405.

Dawson, Hana N., Burlingame, Richard, and Cannons, Andrew C. (1997). Stable Transformation of Chlorella : Rescue of Nitrate Reductase-Deficient Mutants with the Nitrate Reductase Gene. *Current Microbiology*, 35, 356–362.

Dufossé, Laurent, Galaup, Patrick, Yaron, Anina, Arad, Shoshana Malis, Blanc, Philippe, Chidambara Murthy, Kotamballi N., and Ravishankar, Gokare A. (2005). Microorganisms and microalgae as sources of pigments for food use: A scientific oddity or an industrial reality? *Trends in Food Science & Technology*, 16, 389–406.

Geng, D. G., Han, Y., Wang, Y. Q., Wang, P., Zhang, L. M., Li, W. B., and Sun, Y. R. (2004). Construction of a system for the stable expression of foreign genes in Dunaliella salina. *Acta. Bot. Sin.*, 36(3), 342–346.

Ghirardi, M (2000). Microalgae: A green source of renewable H2. *Trends in Biotechnology*, 18, 506–511.

Gilbert, L.A., Larson, M.H., Morsut, L., Liu, Z.R., Brar, G.A., Torres, S.E., Stern-Ginossar, N., Brandman, O., Whitehead, E.H., Doudna, J.A., et al. (2013). CRISPR-mediated modular RNA-guided regulation of transcription in eukaryotes. *Cell*, *154*, 442–451.

Gupta, V., Bijo, A.J., Kumar, M., Reddy, C.R.K., and Jha, B. (2012). Detection of epigenetic variations in the protoplast-derived germlings of ulva reticulata using methylation sensitive amplification polymorphism (MSAP). *Mar. Biotechnol.*, *14*, 692–700. doi: 10.1007/s10126-012-9434-7

Hodge, F.J., Buchanan, J., and Zuccarello, G.C. (2010). Hybridization between the endemic brown algae Carpophyllum maschalocarpum and Carpophyllum angustifolium (Fucales): Genetic and morphological evidence. *Phycol. Res.*, *58*, 239–247. doi: 10.1111/j.1440-1835.2010.00583.x

Horvath, P., and Barrangou, R. (2010). CRISPR/cas, the immune system of bacteria and archaea. *Science*, *327*, 167–170.

Howell, S.H. (1998). *Molecular Genetics of Plant Development*. Cambridge University Press, Cambridge, UK.

Jiang, P., Qin, S., and Tseng, C.K. (2003). Expression of the lacZ reporter gene in sporophytes of the seaweed Laminaria japonica (Phaeophyceae) by gametophyte-targeted transformation. *Plant Cell Rep.*, *21*, 1211–1216.

Jiang, W.Z., Zhou, H.B., Bi, H.H., Fromm, M., Yang, B., and Weeks, D.P. (2013). Demonstration of crispr/cas9/sgrna-mediated targeted gene modification in Arabidopsis, tobacco, sorghum and rice. *Nucleic Acids Res.*, *41*, e188.

Jinek, M., Chylinski, K., Fonfara, I., Hauer, M., Doudna, J.A., and Charpentier, E. (2012). A programmable dual-RNA-guided DNA endonuclease in adaptive bacterial immunity. *Science*, *337*, 816–821.

Kaeppler, S.M., Kaeppler, H.F., and Rhee, Y. (2000). Epigenetic aspects of somaclonal variation in plants. *Plant Mol. Biol.*, 43, 179–188. doi: 10.1023/A:1006423110134.

Keeling, Patrick J., Burger, Gertraud, Durnford, Dion G., Lang, B. Franz, Lee, Robert W., Pearlman, Ronald E., Roger, Andrew J., and Gray, Michael W. (2005). The tree of eukaryotes. *Trends in Ecology & Evolution*, 20, 670–676. doi:10.1016/j.tree.2005.09.005.

Keeling, P.J., Burger, G., Durnford, D.G., Lang, B.F., Lee, R.W., Pearlman, R.E., Roger, A.J., Gray, Oliver, K., Benemann, C.S.E., Niyogi, K.K., and Vick, B. (2011). High-efficiency homologous recombination in the oil-producing alga nannochloropsis Sp. *Proc. Nat. Acad. Sci. USA*, *108*(52), 21265–21269.

Kilian, Oliver, Benemann, Christina S. E., Niyogi, Krishna K., and Vick, Bertrand (2011). High-efficiency homologous recombination in the oil-producing alga *Nannochloropsis* sp. *Proceedings of the National Academy of Sciences*, 108, 21265–21269.

Kirk, D.L. (2003). Seeking the ultimate and proximate causes of volvoxmul- ticellularity and cellular differentiation. *Integr. Comp. Biol. 43*, 247–253. doi: 10.1093/icb/43.2.247

Kito, H., Kunimoto, M., Kamanishi, Y., and Mizukami, Y. (1998). Protoplast fusion between *Monostroma nitidum* and *Porphyra yezoensis* and subsequent growth of hybrid plants. *J. Appl. Phycol.*, *10*, 15–21.doi: 10.1023/A:1008063415548

Kohlmeyer, J. (1975). New clues to the possible origin of ascomycetes. *BioScience*, *25*, 86–93.

Konotchick, T., Dupont, C.L., Valas, R.E., Badger, J.H., and Allen, A.E. (2013). Transcriptomic analysis of metabolic function in the giant kelp, *Macrocystispyrifera*, across depth and season. *New Phytol.*, *198*, 398–407.

Kumar, S.V., Misquitta, R.W., Reddy, V.S., Rao, B.J., and Rajam, M.V. (2004). Genetic transformation of the green alga—Chlamydomonas reinhardtii by agrobacterium tumefaciens. *Plant Sci.*, *166*(3), 731–738.

Lapidot, Miri, Raveh, Dina, Sivan, Alex, Arad, Shoshana Malis, and Shapira, Michal (2002). Stable Chloroplast Transformation of the Unicellular Red Alga*Porphyridium* Species. *Plant Physiology*, 129, 7–12.

Larkin, P.J., and Scowcroft, W.R. (1981). Somaclonal variation—A novel source of variability from cell cultures for plant improvement. *Theor. Appl. Genet.*, *60*, 197–214

Lee, R.E. (2008). *Phycology*. Cambridge University Press, Cambridge, UK, p. 194.

Legall, Y., Brown, S., Marie, D., Mejjad, M., and Kloareg, B. (1993). Quantification of nuclear-DNA and G-C content in marine macroalgae by flow-cytometry of isolated-nuclei. *Protoplasma, 173*, 123–132.

Lerche, Kai, and Hallmann, Armin (2009). Stable nuclear transformation of Gonium pectorale. *BMC Biotechnology*, 9, 64.

Lewin, R.A. (1958). Genetics and marine algae. In Berkeley, L.A., and Buzzati-Traverso, A.A. (Eds.), *Perspectives in Marine Biology* (pp. 547–557). University of California Press, Berkeley, CA.

Li, J.F., Norville, J.E., Aach, J., McCormack, M., Zhang, D.D., Bush, J., Church, G.M., and Sheen, J. (2013). Multiplex and homologous recombination-mediated genome editing in Arabidopsis and Nicotianabenthamiana using guide RNA and Cas9. *Nat. Biotechnol., 31*, 688–691.

Li, T., Liu, B., Spalding, M.H., Weeks, D.P., and Yang, B. (2012). High-efficiency TALEN-based gene editing produces disease-resistant rice. *Nat. Biotechnol., 30*, 390–392.

Lin, Hanzhi, and Qin, Song (2014). Tipping Points in Seaweed Genetic Engineering: Scaling Up Opportunities in the Next Decade. *Marine Drugs*, 12, 3025–3045. doi:10.3390/md12053025.

Liu, W.S., Yuan, J.S., and Stewart, C.N. (2013). Advanced genetic tools for plant biotechnology. *Nat. Rev. Genet., 14*, 781–793.

Lüning, K. (1990 Aug 30). *Seaweeds: Their Environment, Biogeography, and Ecophysiology*. John Wiley & Sons.

Maggs, C.A., Verbruggen, H., and DeClerck, O. (2007). Molecular systematics of red algae: building future structures on firm foundations. In Brodie J. and Lewis J.M. (Eds.), *Unravelling the Algae: The Past, Present and Future of Algal Systematics* (pp. 103–121). Taylorand Francis, Boca Raton, FL; London, UK; New York, NY.

Mahfouz, M.M., Li, L.X., Shamimuzzaman, M., Wibowo, A., Fang, X.Y., and Zhu, J.K. (2011). De novo-engineered transcription activator-like effector (TALE) hybrid nuclease with novel DNA binding specificity creates double-strand breaks. *Proc. Natl. Acad. Sci. USA, 108*, 2623–2628.

Mardis, E.R. (2011). A decade's perspective on DNA sequencing technology. *Nature, 470*, 198–203.

Mardis, E.R. (2013). Next-generation sequencing platforms. *Annu. Rev. Anal. Chem., 6*, 287–303.

Matsumura, W., Yasui, H., and Yamamoto, H. (2000). Mariculture of Laminaria japonica (Laminariales, Phaeophyceae) using protoplast regeneration. *Phycol. Res., 48*, 169–176. doi: 10.1111/j.1440-1835.2000.tb00213.x

Matsumura, W., Yasuj, H., and Yamamoto, H. (2001). Successful sporophyte regeneration from protoplasts of Undariapinnatifida (laminariales, phaeophyceae). *Phycologia, 40*, 10–20. doi: 10.2216/i0031-8884-40-1-10.1

Matsunaga, T., Takeyama, H., and Nakamura, N. (1990). Characterization of cryptic plasmids from marine cyanobacteria and construction of a hybrid plasmid potentially capable of transformation of marine cyanobacterium, Synechococcus Sp., and its transformation. *Appl. Biochem. Biotechnol., 24–25*(1), 151–160.

Matsuo, Y., Imagawa, H., Nishizawa, M., and Shizuri, Y. (2005). Isolation of an algal morphogenesis inducer from a marine bacterium. *Science, 307*, 1598. doi: 10.1126/science.1105486.

Matsuzaki, Ryo, Hara, Yoshiaki, and Nozaki, Hisayoshi (2012). A taxonomic revision of *Chloromonas reticulata* (Volvocales, Chlorophyceae), the type species of the genus *Chloromonas*, based on multigene phylogeny and comparative light and electron microscopy. *Phycologia*, 51, 74–85. doi:10.2216/11-18.1.

Michel, G., Tonon, T., Scornet, D., Cock, J.M., and Kloareg, B. (2010). The cell wall polysaccharide metabolism of the brown alga *Ectocarpussiliculosus*. Insights into the evolution of extracellular matrix polysaccharides in eukaryotes. *New Phytol., 188*, 82–97.

Mehta, Shamir R., Granger, Christopher B., Boden, William E., Steg, Philippe Gabriel, Bassand, Jean-Pierre, Faxon, David P., Afzal, Rizwan, Chrolavicius, Susan, Jolly, Sanjit S., Widimsky, Petr, Avezum, Alvaro, Rupprecht, Hans-Jurgen, Zhu, Jun, Col, Jacques, Natarajan, Madhu K., Horsman, Craig, Fox, Keith A.A., and Yusuf, Salim (2009). Early versus Delayed Invasive Intervention in Acute Coronary Syndromes. *New England Journal of Medicine*, 360, 2165–2175.

Melis, Anastasios (2007). Photosynthetic H2 metabolism in Chlamydomonas reinhardtii (unicellular green algae). *Planta*, 226, 1075–1086.

Meuser, Jonathan E., Ananyev, Gennady, Wittig, Lauren E., Kosourov, Sergey, Ghirardi, Maria L., Seibert, Michael, Dismukes, G. Charles, and Posewitz, Matthew C. (2009). Phenotypic diversity of hydrogen production in chlorophycean algae reflects distinct anaerobic metabolisms. *Journal of Biotechnology*, 142, 21–30.

Miller, S.M., and Kirk, D.L. (1999). glsA, a Volvox gene required for asymmetric division and germcell specification, encodes a chaperone-like protein. *Dev. Camb. Engl., 126*, 649–658.

Minoda, Ayumi, Sakagami, Rei, Yagisawa, Fumi, Kuroiwa, Tsuneyoshi, and Tanaka, Kan (2004). Improvement of culture conditions and evidence for nuclear transformation by homologous recombination in a Red Alga, Cyanidioschyzon merolae 10D. *Plant and Cell Physiology*, 45, 667–671.

Moscou, M.J., and Bogdanove, A.J. (2009). A simple cipher governs DNA recognition by TAL effectors. *Science*, *326*, 1501.

Mussolino, C., Morbitzer, R., Lutge, F., Dannemann, N., Lahaye, T., and Cathomen, T. (2011). A novel tale nucleas scaffold enable shigh genome editing activity in combination with low toxicity. *Nucleic Acids Res.*, *39*, 9283–9293.

Nakamura, Y., Sasaki, N., Kobayashi, M., Ojima, N., Yasuike, M., Shigenobu, Y., Satomi, M., Fukuma,Y., Shiwaku,K., Tsujimoto, A., et al. (2013). The first symbiont-free genome sequence of marine red alga, Susabi-nori (*Pyropia yezoensis*). *PLoS One*, *8*, e57122.

Nekrasov, V., Staskawicz, B., Weigel, D., Jones, J.D.G., and Kamoun, S. (2013). Targeted mutagenesis in the model plant Nicotianabenthamiana using Cas9 RNA-guided endonuclease. *Nat. Biotechnol.*, *31*, 691–693.

Neuhaus, G., Neuhausurl, G., Degroot, E.J., and Schweiger, H.G. (1986). High-yield and stable transformation of the unicellular green-alga acetabularia by microinjection of SV40 and PSV2NEO. *EMBO J*, *5*, 1437–1444.

Neuhaus, G. and Spangenberg, G. (1990). Plant transformation by microinjection techniques. *Physiologia Plantarum*, *79*(1), 213–217.

Neumann, E., Schaefer-Ridder, M., Wang, Y., and Hofschneider, P.H. (1982). Gene transfer into mouse lyoma cells by electroporation in high electric fields. *EMBO J.*, *1*(7), 841–845.

Ngan, Chew Yee, Wong, Chee-Hong, Choi, Cindy, Yoshinaga, Yuko, Louie, Katherine, Jia, Jing, Chen, Cindy, Bowen, Benjamin, Cheng, Haoyu, Leonelli, Lauriebeth, Kuo, Rita, Baran, Richard, García-Cerdán, José G., Pratap, Abhishek, Wang, Mei, Lim, Joanne, Tice, Hope, Daum, Chris, Xu, Jian, Northen, Trent, Visel, Axel, Bristow, James, Niyogi, Krishna K., and Wei, Chia-Lin (2015). Lineage-specific chromatin signatures reveal a regulator of lipid metabolism in microalgae. *Nature Plants*, 1.

Otto, S.P., and Gerstein, A.C. (2008). The evolution of haploidy and diploidy. *Curr. Biol.*, *18*, R1121–R1124.

Pappas, Valeria, and Miller, Stephen M. (2009). Functional analysis of the Volvox carteri asymmetric division protein GlsA. *Mechanisms of Development*, 126, 842–851. doi:10.1016/j.mod.2009.07.007.

Pedersen, P.M. (1981). Phaeophyta: Life histories. In Lobban, C.S. (Ed.), *The Biology of Seaweeds* (pp. 194–217). University of California Press, Berkeley, CA, USA.

Podevin, N., Devos, Y., Davies, H.V., and Nielsen, K.M. (2012). Transgenic or not? No simple answer! *EMBO Rep.*, *13*, 1057–1061.

Polne-Fuller, M., Biniaminov, M., and Gibor, A.(1984). Vegetative propagation of porphyra perforata. *Proc. Int. Seaweed Symp., 11*, 308–313. doi: 10.1007/978-94-009-6560-7_60

Qin, S., Lin, H., and Jiang, P. (2012). Advances in genetic engineering of marine algae, *Biotechnology Advances*, *30*(6), 1602–1613.

Reddy, C.R.K., Migita, S., and Fujita, Y. (1989). Protoplasts isolation and regeneration of three species of Ulva in axenic culture. *Bot. Mar.*, *32*, 483–490. doi: 10.1515/botm.1989.32.5.483.

Reznikoff, William S. (2008). Transposon Tn*5*. *Annual Review of Genetics*, 42, 269–286.

Roessler, P.G., and Ohlrogge, J.B. (1993). Cloning and characterization of the gene that encodes acetyl-coenzyme A carboxylase in the alga Cyclotella cryptica. *Journal of Biological Chemistry*, 268, 19254–19259.

Rusk, N. (2014). CRISPRs and epigenome editing. *Nat. Methods*, *11*, 28–28.

Sessoms, A.H., and Huskey, R.J. (1973). Genetic control of development in Volvox: Isolation and characterization of morphogenetic mutants. *Proc. Natl. Acad. Sci. U.S.A.*, *70*, 1335–1338. doi: 10.1073/pnas.70.5.1335

Shalem, O., Sanjana, N.E., Hartenian, E., Shi, X., Scott, D.A., Mikkelsen, T.S., Heckl, D., Ebert, B.L., Root, D.E., Doench, J.G., et al. (2014). Genome-scale CRISPR-Cas9 knockout screening in human cells. *Science*, *343*, 84–87.

Shan, Q.W., Wang, Y.P., Li, J., Zhang, Y., Chen, K.L., Liang, Z., Zhang, K., Liu, J.X., Xi, J.J., Qiu, J.L., et al. (2013). Targeted genome modification of crop plants using a CRISPR-cas system. *Nat. Biotechnol.*, *31*, 686–688.

Shukla, V.K., Doyon, Y., Miller, J.C., DeKelver, R.C., Moehle, E.A., Worden, S.E., Mitchell, J.C., Arnold, N.L., Gopalan, S., Meng, X.D., et al. (2009). Precise genome modification in the crop species Zea mays using zinc-finger nucleases. *Nature*, *459*, 437–441.

Sizova, I., Greiner, A., Awasthi, M., Kateriya, S., and Hegemann, P. (2013). Nuclear gene targeting in Chlamydomonas using engineered zinc-finger nucleases. *Plant J.*, *73*, 873–882.

Spoerner, M., Wichard, T., Bachhuber, T., Stratmann, J., and Kendal, W.S. (2012). Ulvamutabilis (Chlorophyta) depends on a combination of two bacterial species excreting regulatory factors. *J. Phycol.*, *48*, 1433–1447. doi: 10.1111/j.1529-8817.2012.01231.x

Starr, R.C. (1969). Structure, reproduction and differentiation in Volvox carterif. nagariensisIyengar, HK9 and 10. *Arch. Protistenkd.*, *11*, 204–222.

Steinbrenner, Jens, and Sandmann, Gerhard (2006). Transformation of tHe Green Alga *Haematococcus pluvialis* with a Phytoene Desaturase For Accelerated Astaxanthin Biosynthesis. *Applied and Environmental Microbiology*, 72, 7477–7484.

Sun, Yu, Yang, Zhiyong, Gao, Xiaoshu, Li, Qiyun, Zhang, Qingqi, and Xu, Zhengkai (2005). Expression of Foreign Genes in *Dunaliella* by Electroporation. *Molecular Biotechnology*, 30, 185–192.

Tan, C., Qin, S., Zhang, Q., Jiang, P., and Zhao, F. (2005). Establishment of a micro-particle bombardment transformation system for Dunaliella salina. *J Microbiol*, 43(4), 361–365.

Tanner, C.E. (1981). Chlorophyta: Lifehistories. In Lobban, C.S., and Wynne, M. (Eds.), *The Biology of Seaweeds* (pp. 218–247). University of California Press, Berkeley, CA, USA.

Te, Michael R., Lohuis, and Miller, David J. (1998). Genetic transformation of dinoflagellates (Amphidinium and Symbiodinium): expression of GUS in microalgae using heterologous promoter constructs. *The Plant Journal*, 13, 427–435.

Tolonen, A.C., Liszt, G.B., and Hess, W.R. (2006). Genetic manipulation of Prochlorococcus strain MIT9313: Green fluorescent protein expression from an RSF1010 plasmid and Tn5 transposition. *Appl. Environmental Microbiol.*, *72*(12), 7607–7613.

Tseng, C.K. (1987). Laminaria mariculture in China. In Doty, M.S., Caddy, J.F., and Santelices, B. (Eds.), *Case Studies of Seven Commercial Seaweed Resources* (p. 311). Food and Agriculture Organization of the United Nations, Rome, Italy.

Ueki, N., and Nishii, I. (2009). Controlled enlargement of the glycoprotein vesicle surrounding a Volvox embryo requires the InvB nucleotide-sugar transporter and is required for normal morphogenesis. *Plant Cell*, *21*, 1166–1181. doi: 10.1105/tpc.109.066159

Van derMeer, J.P., and Todd, E.R. (1980). The life history of Palmariapalmata in culture. A new type for the Rhodophyta. *Can. J. Bot.*, *58*, 1250–1256.

Von Stosch, H.A. (1965). The sporophyte of LiagorafarinosaLamour. *Br. Phycol. Bull.*, *2*, 486–496.

Voytas, D.F. (2013). Plant genome engineering with sequence-specific nucleases. *Annu. Rev. Plant Biol.*, *64*, 327–350.

Wang, T., Wei, J.J., Sabatini, D.M., and Lander, E.S. (2014). Genetic screens in human cells using the CRISPR-cas9 system. *Science*, *343*, 80–84.

Weeks, D.P. (2011). Homologous recombination in Nannochloropsis: A powerful tool in an industrially relevant alga. *Proc. Natl. Acad. Sci. USA*, *108*, 20859–20860.

West, J.A., and Hommersand, M.H. (1981). Rhodophyta: Lifehistories. In Lobban, C.S., and Wynne, M. (Eds.), *The Biology of Seaweeds* (pp.133–193). University of California Press: Berkeley, CA, USA.

Wu, J., Du, H., Liao, X., Zhao, Y., Li, L., and Yang, L. (2011). Tn5Transposase-assisted transformation of Indica Rice. *Plant J.*, *68*(1), 186–200.

Xu, Han, Miao, Xiaoling, and Wu, Qingyu (2006). High quality biodiesel production from a microalga Chlorella protothecoides by heterotrophic growth in fermenters. *Journal of Biotechnology*, 126, 499–507.

Xu, T., Li, Y., Van Nostrand, J.D., He, Z., and Zhou, J. (2014). Cas9-based tools for targeted genome editing and transcriptional control. *Appl. Environ. Microbiol.*, *80*, 1544–1552.

Yao, C., and Jun, L. (2013). Chinese scientists sequence genome of kelp, seafood species. Available online: http://english.peopledaily.com.cn/202936/8442860.html (accessed on 25 March 2014).

Zhang, Y., Zhang, F., Li, X.H., Baller, J.A., Qi, Y.P., Starker, C.G., Bogdanove, A.J., and Voytas, D.F. (2013). Transcription activator-like effector nucleases enable efficient plant genome engineering. *Plant Physiol.*, *161*, 20–27.